Animal Communication

A BLAISDELL SCIENTIFIC PAPERBACK

Paul R. Gross, *Brown University*
CONSULTING EDITOR

Animal

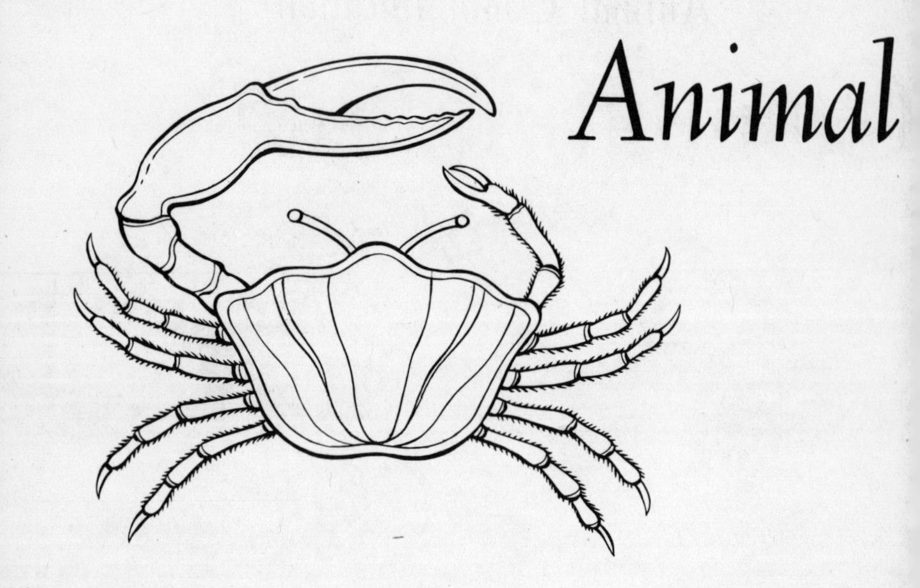

HUBERT and MABLE FRINGS
UNIVERSITY OF HAWAII

Communication

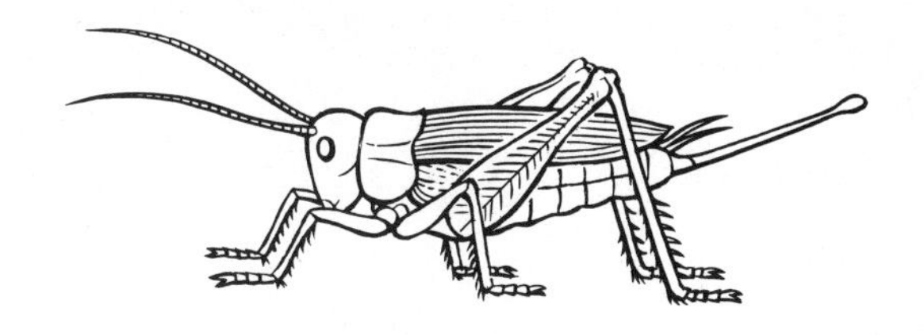

BLAISDELL PUBLISHING COMPANY
A Division of Ginn and Company
NEW YORK · TORONTO · LONDON

FIRST EDITION, 1964

Library of Congress Catalog Card Number 64-15978

TO

Our Animal Associates

FROM WHOM WE HAVE LEARNED SO MUCH

ABOUT WHOM WE KNOW SO LITTLE

Preface

The idea that animals communicate with each other is as old as mankind. Only recently, however, has the study of animal communication emerged as a formal entity in science. The foundations of this study are barely laid; even the definition and limits of the field are still uncertain. This book is an attempt to present some of the concepts that are emerging and to illustrate them with examples drawn from an extensive literature on animal behavior.

This is not meant to be an exhaustive treatise, but is intended primarily for nonspecialists—students, specialists in other fields of science, and educated laymen. With this in mind, literature citations in the text have been avoided, although every effort has been made to verify the facts. Realizing that even biologists might have difficulty in identifying animals by scientific names only, we have tried to relate the species mentioned to larger animal groups at least. Otherwise, some knowledge of biological concepts and terms is assumed. The bibliography, likewise, is not meant to be a research tool, but a list of works which should generally interest nonspecialists. Some references are included for those specialists in animal communication who may find the book useful and want original sources.

We hope that the reader will share our enthusiasm for the study of the remarkable communication systems of animals and will be led to read further or, better yet, to observe for himself. This is a field which can use the talents of professional and amateur alike. With further knowledge, the provisional organization of communication patterns presented here—even possibly many of the "facts"—will undoubtedly need to be revised or discarded. It is stimulating to realize that impending discoveries may soon render the book itself outdated, a brief review of the state of our knowledge—or lack thereof—at this time.

HUBERT AND MABLE FRINGS

Contents

Animal Communication

1

Introduction

HAVE you ever wondered, on seeing a swarm of tiny insects around an outdoor light, how creatures so tiny find mates in a world so vast?

It is hard for us to realize how gigantic this world is for other animals, most of which are so much smaller than we. Many small moths, for instance, cruise in areas some miles in diameter. Suppose we were to convert the moths' travels to human equivalents. The codling moth is about ¾ of an inch long, whereas the trunk of man's body is about 3 feet long. A mile for a codling moth would, therefore, be equal to about 50 miles for man. The male codling moth searches for a mate in areas up to 2 miles in diameter, equal to man's searching an area 100 miles in diameter. Is it any wonder then that these insects have methods for the female to signal to the male so that he can find her?

Honey bees scout up to 2–3 miles from the hive for food sources, thus covering search areas 4–6 miles in diameter. When a scout bee finds a rich source of food, she returns to the hive to recruit others to help gather it. If, again, we state these distances in human equivalents, by comparing the sizes of bee and man, we find that the bee scouting area is equivalent, for man, to an area of 70,000 square miles. This is equal in size to Uruguay or the State of Washington, equivalent to one-third the area of France, and locating an object in it would be a formidable task, even given powers of flight. It would certainly be almost useless for a scout bee to indicate to its hive-mates only that it has found food, without indicating where the food is; thus honey bees have means to communicate this information.

Male and female mosquitoes fly away from the places where they spend their aquatic larval life to hunt for food. Then they must find each other in order to mate. *Anopheles maculipennis*, in Europe, regularly flies 2–3 miles from its breeding place. This is equivalent, for man, to a distance of 400–500 miles. *Aedes vexans*, in Canada, flies as far as 15 miles from the place where it was reared. For a human being, this is equivalent to 3000 miles. A human male seeking a female, under these conditions, would have to search an area of 30,000,000 square miles, equal to one-half the total land mass of the earth. Without some means for the sexes to signal to each other at a distance, the prospects for the species would be dim indeed.

Insects, however, are large compared with many other animals. Some species of moths carry in their ears minute relatives of the spiders, called mites. The ears of moths, obviously, are very tiny, and the mites even more so, for many of them can live in one ear. Consider how vast this world is for a mite, about 1/50 of an inch in length and unable to fly as insects can. Yet somehow males must find females if mating is to occur.

Part of the problem is solved because both sexes are attracted to the same animal, a moth. Once the mites find a moth, they have an area to explore equivalent, in human terms, to a football field. However, a further problem arises. The mites, in feeding, destroy the ear of the moth. Since the moth uses its ears to hear hunting sounds of bats, its major predators, and to avoid them, it is essential that the mites do not invade both ears, and totally deafen the moth. They do not. In a world that for them is almost infinitely vast, these mites find moths to live on, find members of their own species to mate with, and possess systems of communication that allow themselves and the moths to survive.

DEFINITION OF COMMUNICATION

Communication between animals involves the giving off by one individual of some chemical or physical signal, that, on being received by another, influences its behavior.

The definition seems rather straightforward, but there are some difficulties. For instance, is it communication when a human being attracts a mosquito to his warm body? We shall not consider this so, for the mosquito is merely reacting to a feature of the human body over which the human being has no control. One might be tempted to say that the human being does not give off warmth for the purpose of attracting mosquitoes, whereas a female moth gives off an odor for the purpose of attracting a male. This, however, raises the question of purpose, and it is impossible to know whether a female moth has any purpose in giving off an odor. We can avoid the difficulty, however, by considering communication to be involved only when the sender uses some specialized structures or methods to produce signals.

Some persons would use the term "communication" only to refer to signaling by an individual at some distance from another. This, however, is too restrictive. After all, do we not think that communication has occurred when one person squeezes the hand of another to denote sympathy? Why should we, then, deny tactile communication to animals?

Does a flower communicate with bees when it gives off a scent that attracts them? Certainly the bees' behavior is influenced by the scent, and so the scent could be regarded as a chemical signal. However, one might think that the plant has not really produced a signal, because the odor is built right into it. Unfortunately for this idea, there are plants that produce odors only at certain times of the day or night, when insects are available to fertilize them. The yucca plant, for instance, opens its flowers only at night, when yucca moths are active. Certainly, this is a border-line case, but it more nearly resembles the attraction of mosquitoes to a human being by body-heat.

For our purposes, we shall adopt the idea that, in communication, the sender and receiver are of the same species. Where individuals of one species react to signals of another, the reactions are probably learned and usually not shown by all members of the species. This is illustrated by the African Honey Guides. These birds feed on beeswax, but cannot tear open hollow trees containing bee colonies. Therefore, if one of them finds such a bee-tree, it signals its find to a weasel-like mammal, the ratel, or to a human being, by flying overhead and calling. Both, in time,

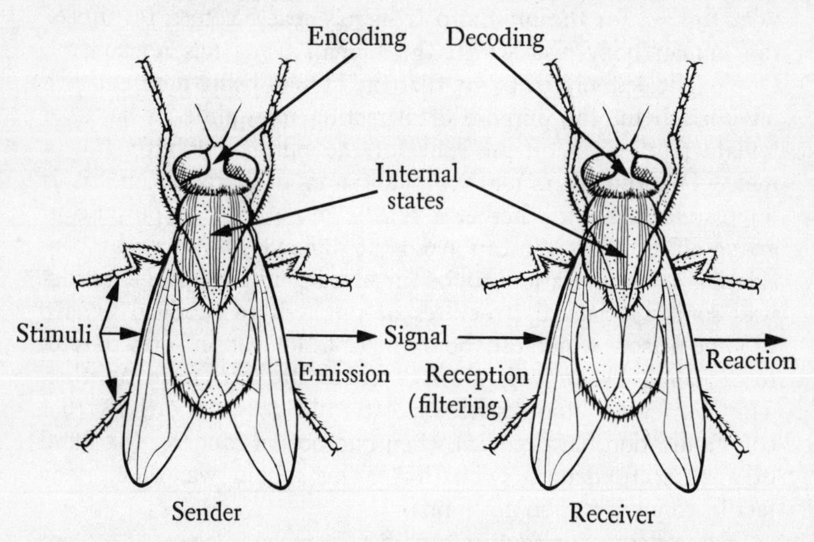

FIGURE 1. *Diagram of animal communication.*

learn what this means and follow the bird to the bee-tree. They then expose the honeycombs, kill the bees, and steal the honey. This leaves the wax for the Honey Guide to eat unmolested.

Our operant definition is illustrated in Figure 1. Here we show an individual, the sender, under influence of external and internal factors, producing a chemical or physical entity, called the signal.

This signal is received through the sense-organs of a receiver, causing changes in internal conditions which result in changes in external behavior. We can study scientifically all phases of this operation: the methods used by the sender to produce the signals, and the influences of internal and external factors on the methods; the chemical or physical nature of the signals; the means by which the signals are received; and the reactions of the receiver, as affected by external and internal factors.

COMMUNICATION IN THE ANIMAL KINGDOM

Communication, thus defined, occurs throughout the Animal Kingdom. Among the simplest of animals, if they be animals at all, are the slime molds. For much of their lives slime molds crawl about feeding on the forest floor as tiny, amoeba-like creatures. They form multicellular groups only to reproduce. An individual gives off a special chemical which attracts others nearby; these crawl to the "caller" and fuse with it to form the breeding body.

At the other end of the scale of complexity in the Animal Kingdom are the elaborate communication systems of honey bees and man. Bees can tell their hive-mates the distance, direction, and nature of food with great precision. Man, as we all know, has the most elaborate and sensitive of all communication systems in the Animal Kingdom, usable even for transmitting information from one generation to the next.

The degree to which animals use communication systems in their daily lives is directly related to the level of development of their sensory-neural systems. Animals, such as jellyfish, that have simple nervous systems, have few channels for communication, and communicate little in general. On the other hand, animals such as insects and vertebrates, that have well-developed senses and highly developed nervous systems, use a multitude of signals for a wide variety of purposes.

Since communication signals must be received through sense-organs, the nature of the sense-organs determines the available communicative channels. The tactile senses and the chemical senses, that is, taste and smell, are well-developed in almost all animals, and are the most widely used of all the senses for communication. However, where the visual sense and auditory sense are highly developed, as in the vertebrates and insects, these are the major communication channels. Visual or auditory signals can communicate a greater quantity of information more exactly than other kinds of signals.

USES OF COMMUNICATION BY ANIMALS

In man, information is mainly transmitted by speaking and writing. Most of this has a social function; some of it has to do with relations between males and females. A small amount is used to transmit information about the surrounding world not immediately related to biological needs, as this book does. This last category is probably missing from the communication systems of animals, for their lives are held too precariously to waste time on it. Otherwise, animals transmit essentially the same major classes of information as do humans.

Man does not have problems in identifying his own species, although he may be concerned about differentiating his particular group or race from others. Animals, however, need to be able to identify other members of their own species, and they have communication signals for this purpose. In some animals, species identification is used to form aggregations, but this is not the usual pattern. Usually, individual animals live more or less separately, except during mating, and they use signals to identify themselves to other members of the species, so that spacing can be maintained, thus assuring a food supply for each individual.

Wherever animals exist in social relationships with each other, communication signals are used for many purposes. Even if animals do not aggregate, they may have warning signals which can be used to alert others. These signals may also serve to ward off prospective predators by calling in cohorts to frighten the enemies. Most animals feed separately, but higher animals share food. In such cases, signals may be used to indicate sources of food. These may be merely attractive, such as those of some birds, or they may evolve into a system of complex guidance signals, such as those which social insects use to guide their fellows from the nest to distant food sources.

By far the most important animal communication signals are used in mating. Males and females must somehow come together, so that the fertilized eggs necessary to found the next generation can be produced. Where males and females of different species are similar in appearance, species identification becomes doubly important. Sperms should not be sent to eggs that they cannot fertilize. Communication signals act as "passwords" to identify members of a species to each other. They also afford a method for the sexes to meet that is safer than mere chance wandering.

When the sexes are brought together, communication signals may be used in courtship and mating. Often the attractive signals have afforded only rudimentary identification. Before actual mating occurs, positive identification must be made. The courtship dances of spiders and rituals of birds serve to identify the males to the females and to prepare both for the final mating act. Generally, mating is a rather exact process, requiring a great deal of cooperation between male and female. Often it involves a reversal of the usual habits: for example, predatory forms, which jump on anything that moves in front of them, must not pounce upon and kill prospective mates.

The mating act results in the production of fertilized eggs and ultimately young. Some animals give no care to either, but many animals care for the eggs or for both the eggs and the young. If both male and female care for the eggs, they must have means of signaling to each other about food or danger. Where the adults take care of the young, as do the birds and mammals, family life develops, and signals used by the adults take on added importance, for the young also respond to them. As the young grow, they develop their own signals, and these affect the behavior of the parents. It is no accident that the animal with the most highly evolved family type of social organization—man—has developed the most elaborate communication system.

WHY STUDY ANIMAL COMMUNICATION?

There are at least three reasons why biologists study animal communication. First, there is the hope that an understanding of communication systems of animals can be used to manage useful species or to control pest species. Second, there is the belief that studies on communication of animals will disclose the biological origins of human communication and suggest methods of communication other than those we now have. Both of these have practical and useful implications for man.

Third, there is the often impractical goal of zoologists: to understand animals better. A long-standing difference of opinion among zoologists divides those who regard animals as totally predictable machines, from those who regard animals as complex, often mysterious, organisms. Studies on animal communication can lead us to sympathize with, and draw on, both of these viewpoints. Certainly, we have every reason to use all the means afforded by chemistry and physics to study the physiology of the signal-sender, the physical and chemical nature of the signals, and the means by which the signals are converted by the sense-organs of the receiver into nerve impulses. When, however, the signals reach the central nervous system of the receiver we find elements of indeterminancy. These need not be due to mysterious forces in the animal. It is just that the human mind, in trying to conceive of the ultimate complexity of this system, is limited. Perhaps one of the difficulties is that our own communication system breaks down when faced with anything so complicated.

The study of animal communication, then, is central for an understanding of animal behavior in general. More and more, biologists are coming to realize that animals cannot be completely understood by knowing only their internal chemistry and physics. Not that these are unimportant, but their ultimate, and at present generally indeterminate, manifestation is the solitary and social behavior of animals. It is this fascinating field that we shall be reviewing here.

2

Animal Communication Mechanisms

Communication signals are chemical or physical in nature. They are sent out into an environment that is already filled with many other chemical and physical entities. Odors are sent out into a world that is full of odors, and sounds into a world that is full of sounds. If animals are to use signals to identify themselves and to transmit specific information, the signals should have one important attribute: they should be highly improbable at the time and place where they are found.

It is highly improbable, in the middle of the night or in the depths of the ocean, that there be a light flashing with regularity. A flashing light is produced as a visual signal by some animals. If a particular sound has a definitely timed structure, for example, a series of regularly repeated chirps, it is improbable that it would arise merely from the wind rustling in the trees. The improbable signals, moreover, must be sent in channels that are available to prospective receivers.

SENSORY CHANNELS USED BY ANIMALS

Almost all the senses are used by animals for communication. Many animals use more than one sensory channel, and some of the higher animals use all, usually combining more than one to be sure that the signal gets through. We shall discuss all the possible signal channels, with some examples of the use of each.

Tactile systems

Through its tactile senses, an animal can detect objects in contact with it. It is difficult, even in man, to separate the different receptors that are located in the skin and that are stimulated by objects touching the body, so we shall not try to do this for other animals. Suffice it to say that, for communication to occur in a tactile channel, individuals must be in contact with each other. Thus, communication in this channel is restricted to very short ranges.

Types of contacts can be widely varied in nature and in timing, so that the signal possibilities—the potential codes—are very great. The signals can be changed rapidly in intensity and nature, and can be readily turned on and off. Tactile signals are, therefore, particularly useful for transmission of quantitative information, where continuously variable signals are needed to transmit continuously variable items of information, such as distance.

The production of tactile signals by the sender is usually brought about by special movements: grasping, using organs such as claws and claspers, or jostling with the whole body, and so on. For instance, honey bees run about in the hive, bumping into their hive mates, to signal to them. The signals are received through the contact receptors on the surface of the animals. These receptors are probably the least studied of all, even in man. Some animals have special tactile receptors of exquisite sensitivity. Honey bees, for example, have receptors for movements in their legs so sensitive that any increase in sensitivity would cause them to detect the movements of particles in the insects' blood. The long "whiskers" around the face of a cat can be moved just the tiniest fraction of an inch before the cat responds. On the other hand, the tactile receptors of other species require large amounts of stimulation before the animals respond.

Communication in tactile channels, as one would expect when these senses are so well-developed in most animals, is found in all animal groups. Some marine Coelenterates exist as colonies of tiny polyps, in which each individual retains an organic connection with all others. If one of these polyps is touched it draws itself together into a ball. At the same time, the other polyps of the group are also notified of the stimulus and contract, even though the stimulus does not reach them directly.

From this simple tactile communication system, we may turn to an elaborate system, such as is found in the honey bee. A bee that has found a source of nectar indicates the location, distance, and value of the find by running in a definite pattern in the hive. The other bees pick up the information by feeling the rhythm and direction of the scout bee's movements. If one were to take only a man-centered view, he might think that the tactile channel is very limited indeed. Study of communication in this channel by the honey bee should quickly disabuse him of this idea. He must also keep in mind that much of the language of love—in man—is tactile.

Chemical systems

The most widely used channels for communication in the Animal Kingdom are the chemical senses. In man, two chemical senses are recognized: smell and taste. In many animals, however, it is difficult or even impossible to separate these accurately. Generally, the distance chemical sense, or sense of smell, is most important in communication.

Chemical signals can act at great distances. They may, however, give little directional information unless there is movement of the water or air containing them. Since, however, there often is such movement, they are widely used by animals to attract members of the opposite sex. In general, there is slow variability to these signals; once an odor is produced, it cannot easily be replaced by another in a patterned time sequence. Odors work far best, to put it another way, as on-off signals: either the odor is produced, or it is not.

Another important characteristic of odors is that they are generally persistent. True, a puff of odor can be made, but generally this persists until it is moved away. Thus, rapid changes in time are difficult. One should not, however, think of this as necessarily disadvantageous, for this persistence can allow an animal to produce a signal that remains for a long period of time without replacement.

Chemical signals may be produced by an animal's body in general, that is, be odors of the skin itself. More usually, if the signals are used in a communication system, as are sexual odors, the chemicals are produced by special glands. These glands elaborate the chemical materials and either secrete them onto the surface of the body, whence they evaporate, or eject the materials into the air or water. In some animals, the glands can be everted suddenly, so that the odors appear suddenly. Some caterpillars, for instance, when alarmed are able to throw out special "horns," which give forth particular scents. The timing of odor production, then, ranges from continuous, as the characteristic species odor, to short bursts, as sexual or alarm odors.

Chemical signals are received through so-called chemoreceptors, organs for taste and smell. Generally, organs for taste are located near the mouth, as one might expect, for they are usually active in the final sampling of food before it is eaten. However, this is by no means always the case; some fish have taste-organs on the fins or tail, and some insects have taste-organs on their antennae or feet. Organs of smell are generally located near the head of an animal, but again this is by no means always true.

In many cases, the organs of taste and smell are closely related, and cannot be clearly distinguished. Aquatic invertebrates tend to be sensitive to chemicals all over the body, and separate organs for taste and smell have not been found. Insects generally have organs of smell on mobile antennae, allowing them to sample odors in "stereo," so to speak. The olfactory organs of insects, however, are not restricted to these parts, for some species have them near the mouth or on other parts of the body. In the vertebrates, the organs of smell are located near the mouth and nose.

Up to this time, studies on taste and smell in animals have yielded few general rules. There seems to be a great deal of variation in responses of these organs to the various chemicals that reach them. In some cases, the organs are rather unspecific and respond to almost any chemical substance. In other cases, the organs are so selective that they respond only to one particular substance. The organs of smell are among the most sensitive known. Male moths of certain species can detect the odor of females with their antennal organs of smell at distances up to 5 miles. For attraction at long distances, this would seem to be the sense of choice.

We shall mention only a few examples of animals that use the chemical senses in communication. Actually, almost every species of animal has its own species scent. Even individual cells may have a species scent, for, when placed in tissue cultures with cells of other species, they can distinguish their fellows.

It is probable that scent is the major channel by which identification of the species occurs in almost all animals. The only large group that might be an exception is the birds, for their sense of smell is believed to be poorly developed. There is no question, however, that individual species of birds have characteristic odors, detectable even to man, whose sense of smell is admittedly poor. Recent studies indicate that the organs of smell in birds are probably more highly developed than was formerly thought. It may be, therefore, that future work will show that birds rely heavily, as do other animals, on chemical signals.

It would be difficult to think of an insect that does not use the sense of smell either for sexual attraction or identification. The same might be said of mammals. From a man-centered point of view, a language of smells might not seem practical, but we should not forget that perfumes are used almost universally by man to heighten sexual interest, and many primitive people can detect friends or enemies by their odor. Most other mammals are strongly oriented toward odors. Anyone who has led a dog on a leash and watched it picking up information with its nose should be convinced of this.

Tastes are not used much in communication. As signals they suffer from the same disadvantages as tactile signals, that is, the animals must be in contact with each other to communicate. In the tree cricket, *Oecanthus*, after the male has attracted the female by a special song, he offers her a fluid produced by a gland on his back; this seems to arouse her to mate. In the giraffe, an important part of the mating process, possibly for recognition of readiness to mate, is the male's tasting of the female's urine.

Optical systems

Whenever animals have well-developed eyes, they develop optical signals. These can work at a distance, but in most cases this distance is not too great. Obviously, the effective distance depends upon the sensitivity of the receptor unit. Optical signals, however, give the best indication of direction. They are rapidly variable in time, and can be used in a great range of intensities, giving them great flexibility. Finally, the signals are transient; they can be turned on and off rapidly, thus allowing precise coding of information.

The most specialized of optical signals are the flashes produced by such luminous organisms as lightning-bugs, or fire-flies. These signals are produced by special photophoric organs, which utilize complex chemical reactions to produce an almost heatless light.

Most animals, however, do not have special systems for light production; after all, these work only in darkness and are, therefore, of limited utility.

All animals have a definite body shape, and many develop specific colors, both of which can act as visual signals. The striking characteristic color patterns of male birds illustrate how a system of species and sex identification can be developed by shape and color. Many animals, however, do not have colors that are specific, yet they can identify each other visually. Such animals have elaborate patterns of movement, different for each species, which transmit the information. It is interesting to note that man also uses these three methods for communication: turning a light on and off at night to produce a coded message; using objects, such as flags, of particular shapes and colors; and using special movements, as in semaphore signaling.

The ability to receive light, particularly the sensitivity to colors or patterns, varies greatly throughout the Animal Kingdom. Except for the arthropods, which include the crabs and insects, and the molluscs, which include the octopuses and squids, invertebrates generally have poorly developed eyes. Most invertebrates have receptors that respond only to light or dark. Some lower vertebrates have eyes that are not much better, but most vertebrates have highly developed eyes, which not only allow distinction between dark and light, but also between shapes and often colors. Each of these different potentialities opens up different avenues of communication. In general it can be said that, where animals have highly developed visual organs, they tend to use these for communication at relatively short range.

The following are a few examples of optical signals used by animals. Many male crabs have specially developed, often brightly colored, claws and legs which they use to threaten rivals or to court females. Male spiders and birds use their appendages in elaborate poses and dance maneuvers to identify their species or sex. Many mammals, such as dogs, apes, or men, use facial expressions, which are simply arrangements of form and color, to signal aggressive tendencies. The whole body of the animal may be used as a visual signal, as in the case of Herring Gulls, whose circling above a food source notifies others that food is there.

All of these are signals that require some light around the animal; they must be used in daylight. Many animals that live in the depths of the ocean or are active only at night have developed special photophoric structures to produce flashes of light, often of characteristic color and usually in some sort of rhythm, and these can be used for signaling.

Acoustical systems

Acoustical signals partake of some of the characteristics of tactile signals, and, since sound is a mechanical movement of a medium, can be received by the tactile organs of some animals. True sounds, however, are received by special organs of hearing, and animals that have highly developed hearing organs rely heavily on sounds for communication. Characteristically, sounds can be received at a distance, often at as great a distance as chemical signals.

With the proper receiving organs, sounds can be quite directional, but the directionality depends upon their nature. This gives them a neat dual utility where animals may need to transmit precise information about their locale at one time, but where this would be best concealed at other times. Sounds have an advantage over light in that they characteristically go around corners, and so are usually not stopped by obstacles. This depends upon the nature of the sounds; sounds of higher frequencies travel much like light. Thus, directionality is better with higher frequencies than with lower frequencies. Sounds can have a great variety of frequencies, intensities, and patterns; there is great potential variability. This variability allows considerable specific differences. Furthermore, sounds can be transient, coming on suddenly and stopping just as suddenly, thus allowing precise timing.

Sounds used by animals for communication are produced in a number of ways. Almost any activity of an animal—chewing, walking, flying—will produce sounds; these may be called incidental sounds. Most of these just contribute to the general noise in the environment and have too little biological improbability to be useful for signaling. However, there are some animals that use these sounds for communicative purposes. For instance, the wing sounds of female mosquitoes attract the males for mating.

Most of the sounds made by insects are percussion sounds, produced by tapping against something, or by scraping comblike structures together. Offhand, one might think that these would be merely noises, with sudden onset, and this is generally the case. However, by using proper rates of movement of combs across teeth, some insects produce rather musical sounds.

The vertebrates, too, have a variety of methods for producing sounds. Some fish vibrate an air-filled bladder; others rub together parts of the skeleton. Mammals use vocal cords vibrated by a moving air-stream; birds use resonating air chambers and columns. The sounds are modified by passing into hollow chambers, or through the mouth which can turn them on and off sharply, causing tremendous variability in the sound patterns. It is understandable that sound signals are very important for most birds and mammals.

Sounds are received by a wide variety of organs. Even in man, sounds can be received as vibrations by the fingers or by the hairs, without the use of ears. Many lower animals receive sounds through what are essentially tactile organs in the skin. In animals that use sounds for communication, however, there usually are special receptors. Male mosquitoes, for instance, use their antennae, which are clothed with fine hairs, the vibrations of which are picked up by a special bulb at the base, called Johnston's organ. Ordinary grasshoppers have their ears on the sides of the body; katydids and crickets have similar ears on their legs. Moths have highly specialized ears on the thorax. These organs can truly be called ears, and, in many insects, they not only act as receivers, but also as filters, separating the important parts of the sounds from background noises. Vertebrates also have special ears which allow differentiation of frequency and temporal patterns of the sound, and thus permit a wide variety of sound signals to be produced for communication.

As examples of animals that use acoustical signals, we may first mention the two groups whose sounds are most familiar: birds and mammals. Many people consider bird songs to be a primitive form of music, and are inclined to believe that they indicate the bird's happiness. As we shall see, bird songs have much more serious purposes. Although we do not call the sounds produced by most mammals singing, we recognize that these sounds are related to their social and sexual life.

Man's ear is poorly adapted to pick up underwater sounds, so it is customary for man to believe that beneath the surface of the sea is "The Silent World," as a popular book called it. As underwater listening devices improve, however, we are beginning to realize that the watery world is by no means silent. Many crabs and lobsters have special organs for the production of sounds. Snapping shrimp (Figure 2) produce explosive cracks, which can be heard at great distances. The sounds produced by barnacles, when opening or closing, and when moving their appendages, can be picked up many miles from beds of barnacles. Many fish, both marine and freshwater, produce sounds, which apparently are used in mating.

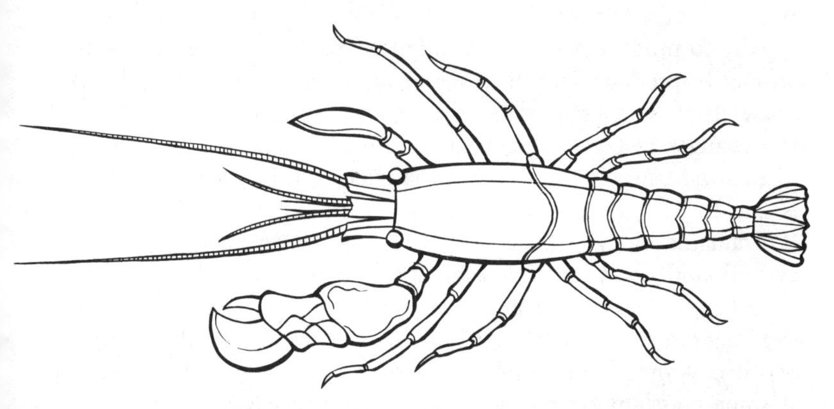

FIGURE 2. *Snapping shrimp. The large claw is the sound-producing device. This marine arthropod lives under rocks near the low tide line. About twice natural size.*

In spring and early summer, the pipings and rattlings of toads and frogs in ponds and streams enliven the evenings. These are sounds produced by males to attract females. In the autumn, in temperate climates, many species of grasshoppers and crickets make the night air alive with a wide variety of songs, again produced by males to attract females.

We must not forget that sound is really rhythmic changes of density in air, water, or even a solid, not just the movements of air which our ears receive. Sounds can, therefore, be carried through solids and liquids in the form of vibrations, and these can be received by animals, and so used for communication. Male spiders, for instance, approaching prospective sexual partners in their webs, must pluck the webs in just the right rhythms, if the females are to welcome them as mates, rather than as food.

Other sensory possibilities

We know very little about the possible use of the thermal sense in communication. Obviously, this could not be used by most aquatic animals, for they are incapable of changing the body temperature to produce a signal. It might also be said that it is not easy for a terrestrial animal to change the body temperature rapidly to produce a signal. It might be possible, however, for an animal to produce heat at a higher level than usual, and thus to make itself obvious. Perhaps the nearest thing to this would be the changes in body temperature of mammals during the phases of the oestrus cycle. In almost all languages, the acceptant female is said to be "in heat," because the body temperature, or at least the radiated energy, seems higher. It has yet to be shown, however, that this acts as an actual signal to the male.

Although insects generally have about the same temperature as the environment, many can produce body temperatures well above environmental, either by warming themselves in the sun, as some grasshoppers do, or by warming themselves through production of heat before and during flight, as some moths do. We know from studies on attraction of mosquitoes to man that insects can receive radiant heat, so the potential receptors are there. It remains to be shown, however, that heat signals are actually used by insects.

The wide distribution of heat receptors in the Animal Kingdom means that a potential communication channel exists. Perhaps, as in the case of the "silent sea," we are just waiting for the correct tools to make the discoveries.

Occasionally, someone suggests that animals can receive energy that man cannot: magnetic lines of force, or radio waves, for instance. We know that some animals, such as honey bees, see ultraviolet light, but this merely means that they can receive a type of energy that we do, but in a different range. As for the other forms of energy, until someone shows that animals can receive them, and that signals and senders actually use them, we cannot speculate.

The same is true for the so-called extra-sensory perception that some psychologists believe humans can use to transmit information. The means of transmission and the nature of the signals, however, are unknown, and thus many scientists doubt that ESP actually exists. As long as we cannot specify the nature of a signal, nor find overt behavior in a sender, it is impossible to test for this in animals scientifically. We should not introduce this idea into our studies, at any rate, unless all other explanations are unreasonable. So far this state of affairs has not arisen.

Given the many channels for communication open to animals, we might ask what sort of information they communicate in them. Before answering this question, however, we shall ask how biologists study the behavior and senses involved in animal communication.

3

Methods of Study

THE study of animal communication is often frustrating, because the animals cannot communicate directly with man. In short, animals cannot talk.

We can only decide whether animals have communicated with each other, or not, by observing their behavior. It is natural to assume that, if some special signal—a sound, an odor, a flash of light, or a special display—is produced, communication is occurring. However, only by careful observation of the behavior of the animals can we decide if communication is taking place and what, if anything, is being communicated.

The study of communication in man may look easy, but this is far from true. Many psychiatrists believe that a major cause of mental illness is a breakdown in communication between the patient and the people around him. Furthermore, it is perfectly obvious that sights, sounds, and smells affect different people in different ways. Thus, even with man, the study can be difficult.

Two major methods are used in studying animal communication: observational methods, in which we observe the animals closely and set up what could be called behavioral inventories, and experimental methods, in which the sender or receiver is interfered with, or operated upon, or the signal is manipulated. In most studies on communication, an investigator is likely to use various combinations of these methods.

OBSERVATIONAL METHODS

For convenience, we may divide observational studies into those on the behavior of the sender, on the nature of the signal, and on the behavior of the receiver. The observation may be made either in the laboratory or in nature. Natural conditions, of course, are best if we wish to discover the use of communication in the lives of the animals. However, we cannot always keep track of individual animals to follow their behavior in the wild. Therefore, laboratory observations are often needed for clarification.

Study of behavior of signal-senders

Every animal communication signal is produced by some structure of the animal body. One might think that an odor emanating from the body as a whole might be an exception, but it is not, for it is produced by glands in the skin. In most cases, odor signals are produced by glands with special structures. Sounds are almost always produced by special structures, and studies of these structures are of great importance in helping us to understand the types of signals that the animals make. Where color patterns are used by animals as signals, we are interested in how the colors are produced and in the origin and development of the patterns.

The hope, in these studies, is to relate the structure of the signal-producing apparatus to its function. With most lower animals we are just at the earliest stages of this research. In spite of the fact that the production of sounds by grasshoppers and crickets, by movements of legs and wings with their stridulating areas, has been observed for years, little or nothing is known about how these structures produce the sounds and how their form is related to the sounds produced. Stridulating organs that appear quite similar, such as those of katydids and tree crickets, produce entirely different sound patterns, and we do not know how. The action of the vertebrate vocal apparatus is much better understood, although the peculiar vocal apparatus of birds, which do not have vocal cords, is still poorly studied.

Much work remains to be done before we shall be able to relate structure to function in most animal groups. The same might be said for attempts to explain how particular types of displays, such as wing flashing, or special patterns of sound, are produced. These actions require coded patterns in the central nervous system producing impulses to the muscles and resulting ultimately in the behavior patterns.

Many communication signals are related to particular functions, and occur only at certain times in the life of the animal. For example, immature frogs and toads are tadpoles, and swim around in the water; the adults live on land. As a consequence, these animals have totally different signals in immaturity and adulthood. Even if the change in body form is not so dramatic, as in birds or mammals, production of signals may change dramatically when maturity is reached. Only adult birds sing. Likewise, only adult grasshoppers and crickets sing; in this case, sound is produced by the wings, and only adults have wings.

The communication signals used in reproduction are often restricted to one sex. In some insects, for instance, only males have sound-producing structures; in others, both males and females can produce sounds, but only one usually does. Thus, sex and sexual maturity of the animal can be important factors.

The sender's behavior is influenced not only by internal factors, but also by external factors. Many birds and insects produce signals only seasonally. Even some marine invertebrates are influenced by the seasons, as reflected in the phases of moon and tides, and mate only at certain times of the year. It is only during the mating seasons that sexual signals are produced; between times the adults behave more or less as do immature animals.

Animals, such as insects, that do not have constant body temperatures, have their behavior seriously affected by external temperatures, for these determine the internal temperatures. Grasshoppers, for instance, have minimum and maximum temperatures below and above which they do not sing. These are by no means the freezing or killing temperatures, either. In between these two extremes, the animals sing at greater speeds as the temperatures are higher. A grasshopper of the eastern United States, *Neoconocephalus ensiger*, doubles its rate of singing for every 10°C rise in temperature. The tape recorder, by allowing us to record sound production at a variety of temperatures, and later to time it accurately and make visible patterns for measurements, enables us to make accurate studies of insect singing rates. These studies, however, are just in their infancy.

The presence or absence of other individuals of the same species near the sender can also influence signal production. Birds and crabs do not give threat signals unless a rival is nearby. Many insects sing either synchronously or alternately with those around them. Male snowy tree crickets, *Oecanthus niveus*, for instance, join in almost synchronous choruses. Many tropical fireflies flash synchronously, apparently following some leader. On the other hand, tree frogs form groups of three, each individual of which alternates with the other two in singing. Katydids sing alternately with each other; if one interrupts the other by starting too soon, the duet is disrupted. Two individuals singing together at the same temperature produce more notes per unit time than one singing by itself. Thus, animal social relations may drastically affect signal production.

The signals themselves may influence the senders. For example, the female silkworm moth cannot smell her own attractive odor; only the male can smell this. This leaves her receptors open to receive other information. Male cicadas have a special muscle that folds the tympanum of the ear, rendering it inactive when they start to sing, thus protecting the ear from the high intensity sound. It is quite possible, from what we know of the structure of the hearing organs of male grasshoppers, that these males may be almost deaf to the sounds they produce. However, the singing males definitely do listen for reflections of the sounds coming back from the surroundings. Birds and mammals can hear their own voices through bone conduction, and for man, at least, we know that this feedback is important in speech.

Study of signals

Information transmitted from one individual to another is coded into a signal, and the coding varies from animal to animal. The English zoologist, Haskell, has pointed out that most of our so-called "dumb animals" are really not dumb at all, they have perfectly good voices, or at least channels of communication. It is just that in many cases man fails to appreciate their means of communication. In some cases, as with grasshoppers that use ultrasonic sounds, man cannot even hear what they produce. Even if we restrict our definition of animal language to communication by sounds alone, many more animals than we once thought have languages.

Tactile signals usually involve some sort of motion, often complex and rapid. Honey bees, for instance, in their communicatory dances, move about on the honeycomb so rapidly that it is difficult for the unaided eye to follow them. Here the movie camera and still camera have proved to be invaluable tools for study. By filming at high speeds, with later playback in slow motion, one can determine the exact nature of the bee dances, which otherwise are almost too rapid to follow.

The analysis of optical signals, if it is to be done accurately, is best done by physicists trained in optics. Light may vary in a number of ways: frequency, which man perceives as color; intensity, which man perceives as brightness; spectral composition, which man perceives as color mixtures or intermediates; and temporal and spatial patterns, which man perceives as movement and form. Physical equipment is available to analyze all these.

The color patterns of animals are produced either by the structure of the colored parts, or by special pigments deposited in the skin. Many colors in the feathers of birds or wings of butterflies are produced by fine grooves; others are produced by special pigments as are those of the hair of mammals. For the analysis of these colored materials, the chemist must be consulted, and many new methods for chemical analysis are now being used to isolate and identify the pigments. Optical signals of animals usually do not depend so much upon intensity or color as upon movement. Here, as in the analysis of tactile signals, a movie camera is needed.

The chemist also enters the scene to study chemical signals used by animals. The methods for their identification and study are like those for identification and study of almost any other compound. First, the chemicals must be extracted from the material in which they exist, next they must be purified, and finally they must be analyzed. After this, if possible, they may be synthesized and the synthetic materials tested to see whether they are as effective as the natural ones. For example, female cockroaches and gypsy moths give off chemical materials that attract males from a great distance. These materials have been extracted from the females, purified, and tested for their effects on the males. Concentrated signals, so to speak, are thus produced; these are many times more effective than natural ones. Actually, manufacturing chemists have long used similar methods for the extraction of compounds, such as musk, from mammalian attractive glands to use in perfumes. The determination of the exact chemical structure of these particular compounds is a job for the chemist rather than for the biologist.

Acoustical signals, like optical signals, are best analyzed by physicists trained, in this case, in acoustics. Although physically sound and light are quite different, they have approximately the same characteristics: frequency, which to man gives them their pitch; intensity, which to man gives them their loudness; frequency composition, which to man gives them their harmonic or disharmonic qualities; and temporal pattern, which gives them their rhythm.

Different species of animals produce sounds that differ in one or all of these features. For instance, two species of closely related tree crickets, living in the same habitat, have songs that differ in their frequency, so that they are distinguishable even to the human ear. Some birds call at low intensities near their nests, thus not attracting predators, but call at high intensities away from their nests. Closely related species of toads produce notes that are of approximately the same frequency, but differ in pulse-rates, thus giving them their distinctive characters.

The analysis of acoustical signals has been tremendously facilitated by the development of the tape recorder, which allows one to store the sounds for future work. From these, the sounds can be converted into light patterns with an oscilloscope or sound spectrograph. The visual patterns are easy to use for determination of characteristics of the sound. In this field, as in the case of optical signals produced by animals, the major work is in the future.

Study of behavior of receiver

The behavior of the receiver is susceptible to study, as is the behavior of the sender, and usually by similar methods. First, one may be interested in how signals are received. This leads to study of the structure and function of the sense-organs of animals, and is a branch of physiology.

Although the structure of the sense-organs of vertebrates, at least of mammals, is now fairly well-known, the structure of the receptors of other animals is often poorly known. For instance, the exact locations of the organs of taste and smell of many common insects are still matters of debate or sheer conjecture. Obviously, until we know where they are, we cannot study their structure. The relationship of structure to function is also a pressing physiological problem. Even for the sense-organs of man, except possibly the ear and eye, considerably more work is needed to clarify these relationships.

In trying to understand communication, we are often interested in the range of sensitivity of the sense-organs; for instance, can the eyes of birds receive the same frequencies and intensities of light as do the eyes of man? Obviously this is going to determine the intensity of signal needed, and the distance at which a signal can be effective. One measure of sensitivity is called the threshold of response, that is, the least amount of stimulating energy or material that can affect the particular sense-organ. Thresholds for many sense-organs, even those of man, are not well-known, and, for large groups of animals, such as insects, are almost unstudied.

Research on the senses of animals may lead one to a study of behavior or communication indirectly. An animal cannot talk,

Innate reaction to stimulus
(male moths attracted to female in cage)

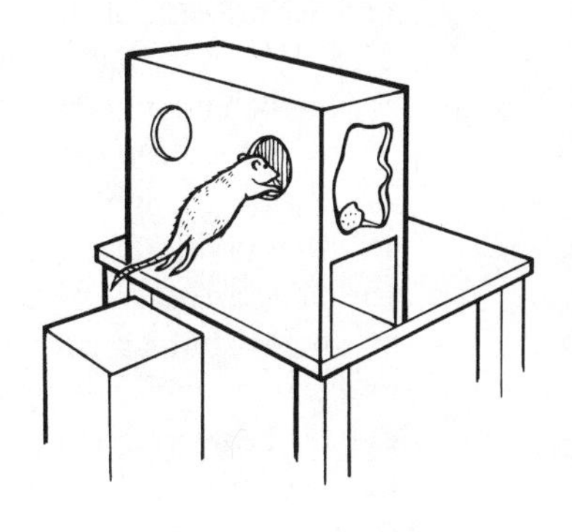

Training reaction to stimulus.
Rat trained to jump at colored circle
to get food hidden behind it

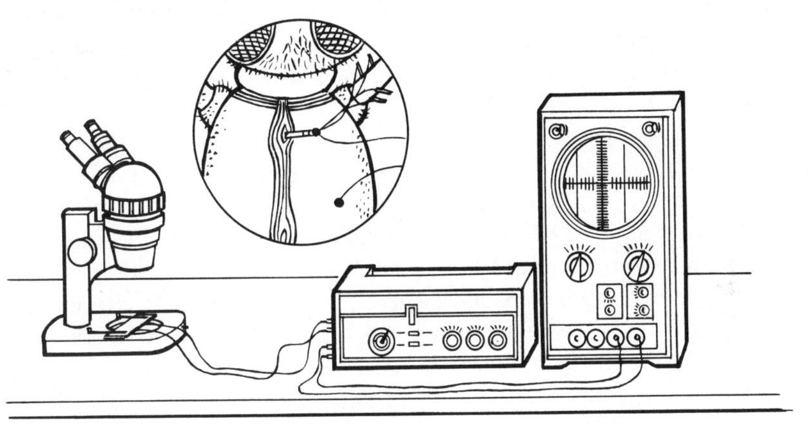

Microscope Amp. Oscilloscope

FIGURE 3. *Methods for studying reception of stimuli by animals.* Upper left: *use of normal behavior of animals (reaction of male moths to odor of female)*; upper right: *use of training (rat trained to jump at colored door to get food)*; lower: *use of electrophysiological methods (insert shows electrodes on nerve-cord of insect under microscope; electrical changes in the cord are amplified and displayed on the oscilloscope screen at the right).*

and so we cannot ask it whether it can receive a particular stimulus. We must have some way of finding out, and we can usually find a behavior pattern, often related to communication, to use as an index of reception. For example, if a fly maggot is placed in a beam of light, it crawls directly away from the light source. If other lights, with various intensities and at various angles to the first, are turned on, the maggot moves from its original path at angles that are determined by the ratios of intensities. Thus, one can study the effect of light on the receptor organs. Obviously, one is also studying the behavior of the maggots. Male cockroaches vibrate their wings when they are stimulated by the odor of females. This reaction can be used to study the receptors involved, and at the same time to study the behavior itself.

A second method to study reception, and one that avoids the use of behavior, is the electro-physiological method. In this method, electrodes are attached to nerves leading from sense-organs, and the electrical potentials produced in the nerves by the activity of the sense-organs are detected by sensitive equipment. This method has yielded a gratifying harvest of information about the basic physiology of the sense-organs. It has been of particular value in the study of the visual and auditory organs of birds, mammals, and insects. Unfortunately, the information obtained may only indicate the potential, rather than the actual, utility to the animal. That a receptor responds to a given stimulus shows only a potential response characteristic of the organ. Whether this particular response of the sense-organ has any behavioral result in an intact animal must be determined by observation of the intact animal. For instance, taste-organs on the antennae of certain butterflies and moths produce nerve potentials in the antennal nerves when sugar is applied to the antennae. The insects react to the sugar, however, by uncoiling their feeding organ, only if all the other taste-receptors, which are on the legs, are completely removed. This result could not be predicted from the electro-physiological study alone.

A third method for studying reception is to train the animal to do something it does not ordinarily do when a stimulus is presented. Quail, to cite a case, learn to come for food to a container carrying an odor that otherwise has no effect on them, if one feeds them only at that container. They then come to any container with the odor, even though no food is there. Similarly, bees can be trained to come to dishes containing sugar-water on which an odor has been placed, and they react to the odor when the sugar-water is removed; this reaction can be used to study their range of sensitivity. This is a behavioral method of study, but, in this case, we obtain insight into the ability of animals to change their behavior patterns as a result of experience.

Obviously, the receptive capacities of an animal determine the types of signals that can be used for communication. A receptor, such as the ear of a noctuid moth, which is sensitive only to sounds above 15,000 cycles per second, cannot receive the sounds produced by the whirring of the moth's wings. However, the relationships between the functional capacities of receptors and the effectiveness of communication may be even more subtle. For instance, the song of the male cricket has a sound frequency of 8000 cycles per second, and this characteristic is most noticeable to man. However, this frequency merely stimulates the female cricket's ear most effectively. The individual pulses of the male's song, about thirty per second, which the male cricket produces by the back and forth movements of his wings, trigger synchronous pulses in the female's nerves, and it is the timing of these pulses that determines her reaction. This receptor, therefore, acts like a radio receiver, filtering out the carrier frequency of the signal and passing on only the "audio" frequency.

The effects of communication signals on the behavior of receivers can be studied by using movies and television to record the visible actions of the animal, and these can later be played back, often at slow motion, for analysis. Anyone who has watched an event under natural conditions, and then later watched the same event again and again in a filmed version, appreciates how much more can be seen during the reshowing of a film. Up to now, the use of these methods has been restricted, because they are expensive. Magnetic tape television recorders, with which one can use tapes over and over, as in sound recording, should soon be available at prices that will allow students of animal behavior to use them, thus facilitating studies of long duration.

The signal receivers, as the senders, are influenced by physiological and motivational states. Generally, immature animals do not respond to sexual signals. For instance, immature female crickets, in the company of singing males, pay no attention to the males' sounds; their ears are not functional until they mature. The sexual condition of mature individuals determines their responses to sexual signals. Female grasshoppers of some species, after they are fertilized and start producing eggs, lose their hear-

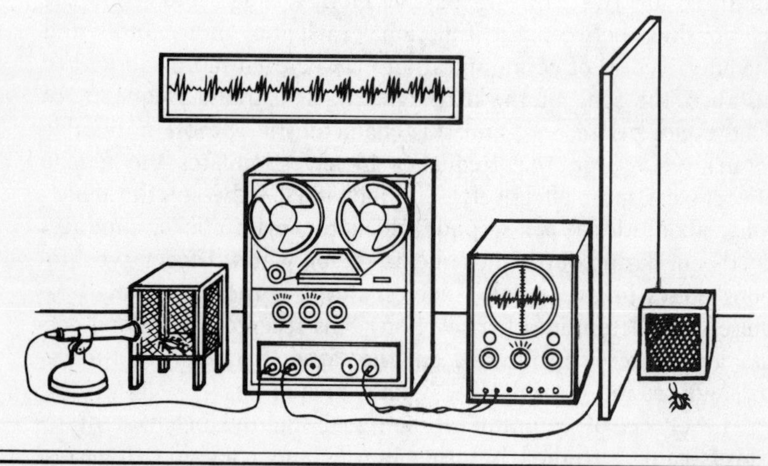

FIGURE 4. *Procedures for studying communication by sounds. Song of insect (left) is recorded; visual pattern of sound is displayed on oscilloscope screen, from which it can be photographed, as in the inset; recorded sound is played to the reacting insect (right).*

ing because the ovaries, swollen with eggs, crush out the air-sacs needed for their ears to act. Thus they no longer can hear the singing males, and they do not waste time reacting to them.

The environment also can affect the receiver's behavior. The behavior of both male and female birds changes sharply with the seasons. At certain seasons the males sing and are ready to mate; at the same time, the females are sexually receptive and come to the males. As we have noted, male crickets and grasshoppers chirp at different speeds at different temperatures. The females, in these species, respond only to the proper speed of singing for their temperature; a female at 25°C does not respond to the singing of a male of the species at 15°C. The presence or actions of other individuals around a receiver may also influence its behavior. To mention just one example, a male deer or elk displays for a female much more vigorously when a rival is present than when he is alone.

EXPERIMENTAL METHODS

The observational methods discussed so far have yielded a great deal of information about the natural communicative behavior of animals. Once we know this, experimental methods can be used to analyze the situation further. Experiments need not be conducted in the laboratory, but can be done in the wild as well. In an experiment, after all, man intrudes himself into the natural situation by changing some condition to see what this does to the behavior of the animals.

Use of synthetic signals

A simple method to change behavior experimentally is to present synthetic signals to the receiver or sender. For instance, if one vibrates the hive of honey bees with sounds of certain frequencies, the bees are rendered inactive. This suggests that vibrations such as these are used in communication by the bees. By imitating the sounds of male katydids with his mouth, one can induce the insects to sing alternately with him, and even to change the number of syllables in their songs when he changes his.

The reactions of animals to synthetic chemicals give us clues to their receptive capacities and also allow us to study their reactions. Synthetic signals can be used most effectively if we know the signals to which an animal responds. We can, however, also use these signals as probes to discover systems of animal communication about which we have no previous knowledge.

Use of "dummies"

Dummies, or decoys, are widely used to study animal communication. In this case, knowing the nature of the signal, we try to imitate it, with modifications, to determine the effects upon the behavior of the animals. We must be cautious in our interpretations, keeping in mind that what seems to be the same signal to us may not be the same to the animal. For instance, a German investigator who was studying the reactions of female crickets to imitations of the chirps of males found that the females did not respond to chirps that sounded perfectly realistic to him and his associates. At the same time, the females insisted, exasperatingly, on responding to chirps transmitted over a telephone which so altered them that, to the human ear, they did not sound at all natural. The explanation was simple, once he unearthed the differences betwen human and cricket hearing. The cricket's ear does not send to the brain information about the frequency of sound, which is the most striking factor for man; instead, the ear sends the tempo of the sound pulses, which is too fast for man to distinguish. As a consequence, the artificial sound used by the investigator, which did not have the proper timing of the pulses, even though its frequency was right, did not move the females. On the other hand, sounds coming through the telephone, which were incorrect in frequency, but had the proper number of pulses, were meaningful to the females.

Dummies of many types are used to study communication. Chemical dummies can be produced by extracting the chemical from the sender and putting this on artificial shapes, or on other animals. The odor of a female mouse that is ready for mating, if applied to an immature male, causes adult males to act toward the immature male as if he were a female. The odor of bee stings excites honey bees in a hive. If this odor is absorbed in wax, and the wax is placed on the landing platform of a beehive, the bees become highly excited, just as if a bee that had stung an enemy were there giving off the scent.

The most familiar dummies or decoys are visual. In a study of the behavior of a fish, the stickleback, investigators used various fishlike shapes. The male of this fish, when in breeding condition, has a bright red belly. This configuration causes other males at their nests to attack. Any dummy that is fishlike, even grotesquely so, and has a red under-side, is threatened by a male stickleback at its nest as if it were another male. Baby Herring Gulls peck at a red spot on the beak of the parents to get food. Almost any pointed object with a red end can be used as a dummy to induce the baby birds to beg food from it, as if it were the parent's beak.

Only recently have we been able to use practical acoustical dummies; the tape recorder has made this possible. Recorded communication signals can now be played back to animals and their behavior observed. For example, when a recording of the call used by Herring Gulls to announce the presence of food is broadcast to Herring Gulls in the air, they gather around the loud speaker in anticipation of food. Similarly, a recording of the chirping of a male tree frog, played through a loud speaker, attracts females to the speaker. As with visual dummies, the acoustical dummies can be altered by an experimenter. We can play a recording backward, or cut it into pieces and play the pieces in different order, or change the frequency or intensity of the sound. By these manipulations, one can discover the effective elements of the signals.

Interference with communication systems

In addition to changing the nature of animals' signals, one can experimentally alter structures or activities of the senders and receivers. One can perform operations upon the animals, so that they no longer can produce or receive signals as they should. For instance, in a study of the relationship between the structure of the stridulatory organ of a male grasshopper and the sounds it produces, an investigator removed teeth from the sound-producing organ. In contrast with what he expected, this had little or no effect on the sounds of the male. In fact, even if he removed almost all the teeth, the sounds the grasshopper produced were almost unaffected, except for loss of intensity; a surprising result. On the other hand, if one removes the large claw that a male fiddler crab uses for aggressive display, the crab loses its dominant position in a group. Male parakeets are identified as such by females, because they have a specially colored patch on the face. When females are painted to match this, the males, which formerly mated with them, are induced to attack them as if they were males.

Other methods

A common method used to find receptors involved in communication is to destroy the suspected sense-organ. The following experiments illustrate this. When the ears on the front legs of female crickets are destroyed, the insects no longer respond to the singing of males. Removal of the antennae of a honey bee renders it incapable of distinguishing members of its own hive from those of others. Since we know that this is done by the sense of smell, the antennae must bear the organs of smell.

In addition to rather direct tampering with communication systems, we can do other, more complicated, experiments. Barriers to the behavior of the receivers can be set up to test the strength of the drive produced in the receiver by the signals. The nature of visual signals and the influence of the environment upon the signaler can be tested by mirror experiments. Fighting fish, for example, display their brilliant colors to their images in mirrors, as if these were rivals. Birds answer recordings of their own voices, when their songs are played back to them. In a sense, this is reversed in experiments in which animals are completely isolated from contact with any other members of the species. Birds have been reared in this way, without contact with their fellows, by Thorpe and his students in England. These investigators have been able by these experiments, to discover what parts of the birds' songs and calls are inborn and what parts are learned from listening to other birds.

As biologists, therefore, we have at our disposal a considerable battery of methods to study the communication systems of animals. With about 700,000 species of animals to study, however, the field is bewilderingly extensive, and we are only beginning to appreciate how much information animals actually transmit by their signals. We shall now review what has been discovered about this so far.

4

Species Identification

MOST animals live in some relation to others of their species. They may live in aggregations, ranging from loose gatherings to complex social organizations; or they may live dispersed, each individual with its particular home-range and feeding area. In either case, they need some means to identify others of the same species, and this is almost always done by communication signals.

AGGREGATIONAL SYSTEMS

Cellular aggregations

The tendency to aggregate is apparently a property of living cells themselves. In tissue cultures, cells from the same animal aggregate and exclude cells from other species. This is presumably brought about by chemical differences which the cells are able to receive. This may not seem to be comparable with communication between whole animals, but it is hard to distinguish from a type of communication found in undoubted whole organisms, the slime molds.

The slime molds are often thought of as part animal and part plant, and therefore not entirely either. For part of their lives, the slime molds are amoeba-like creatures crawling among fallen leaves on the forest floor. When the time comes for reproduction, however, they fuse into a multicellular mass. At that time, one individual starts to puff out a particular chemical, named acrasin, which attracts other individuals of the same species to fuse into the mass. Specific identification seems to come about by the rhythm of puffing of the acrasin. The material is rather rapidly destroyed, and so chemical pulses can be produced in a definite pattern.

Sessile aggregations

Simple aggregations, in which individuals of the same species form groups at approximately the same spot, are found throughout the Animal Kingdom. Sessile species, that is animals that cannot move from their place of attachment, are representative. These organisms almost always have a motile stage in their life, and this does the aggregating. These aggregations have biological value; when the time comes for reproduction, the sexes, which as adults cannot move, are so close that they can easily exchange sperms and eggs.

Two marine animals in which this occurs are oysters and barnacles. Larval oysters, after a time spent in swimming, settle down with members of their own species. They probably use a chemical sense to detect their fellows, but this is not positively known. Barnacles (Figure 5) form large aggregations on rocks or pilings, usually with only one species at one place. Again, it is the larvae which swim about and find members of the same species. If, after cruising around, they cannot find members of their own species, they may be driven to attach themselves near or to members of another species. If they do this, they usually attach right on top or in no apparently significant relation to them. Again, a chemical given off by the attached barnacles is probably the attractant for the larvae.

It may be, however, that the larvae are attracted by vibrations in the water. It is known that barnacles produce sounds when opening or closing, and probably when moving their feeding

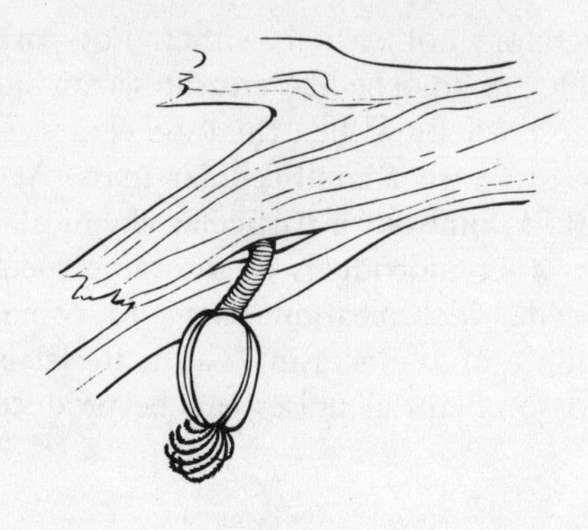

FIGURE 5. *Stalked barnacle (top) and sessile barnacle (bottom). About natural size. These animals feed by catching food with their appendages.*

appendages. The barnacle larvae are well-equipped with vibration-receptors, and therefore could pick up these signals. One might wonder, since barnacles are so small, whether their sounds could travel far enough to be useful in orientation. A recent study in France has shown that sounds from barnacles are still "audible" to underwater microphones as far away as 10 miles from a large bed of the animals. This distance is certainly great enough for use in communication.

Mobile aggregations

Animals that do not remain attached for life may also form aggregations, usually called flocks, swarms, or schools. For example, migratory locusts, the devastating grasshoppers, form large flying swarms which maintain their integrity in flight. It seems likely that locust swarms are held together by sight, smell, hearing, or some combination of these, but we do not know how. However, hearing does not seem to be important, for flying locusts do not respond to broadcasts of the sound of their flight, nor are they disoriented by loud noises.

Fish swim in schools, usually with only one species present. The senses most likely involved are the visual and chemical senses. Fish have well-developed eyes and can distinguish the patterns of other fish. These patterns form signals which could be used for identification. We know that, at least during sexual displays, the visual sense of fish is important. However, we also know that fish produce specific chemicals, which also could be used to maintain the schools. Not only do fish of the same species have the same odor, but fish from the same region sometimes have a characteristic odor. Even individual fish, in some species, have odors that can be distinguished by other members of the species.

Many birds form flocks, with the necessary identification made chiefly by visual and auditory cues. One has only to watch Starlings, as they come to a roost through winter evening skies, to see how integrated and coordinated a flock of birds can be. It is almost certain, since they are silent, that they follow each other with their eyes. It is not obvious whether they have a leader or not, but all seem to turn as if the flock were one individual—a remarkable case of coordination. There must be some signals guiding this behavior, but so far these have not been discovered. When birds fly at night, as they do during migration, special calls are used to keep the flocks together and to recruit grounded individuals. A familiar sound during spring and fall, where geese or ducks migrate, is the honking or quacking as the flocks fly overhead at night; undoubtedly the sounds help to keep these birds together.

Ground-inhabiting birds keep in contact with their fellows mostly by visual signals, embodied in movements and body patterns. Quail and partridge, however, have special calls, the covey calls or rally calls, which are used to maintain flock cohesion. A bird that is separated from the flock, on hearing the call, stops what he is doing and tries to rejoin the flock.

Birds are usually believed to have a poor sense of smell, but this may be incorrect. Quail, for instance, have a body odor which, in captivity, they can transfer to food containers. The quail then give preferential attention to the marked feeders. The characteristic odors of many species of birds, such as albatrosses and petrels, might be a means of species identification. Many of the sea birds that have distinctive odors are active at night, and the replacement of visual signs with chemical signs could be of considerable importance.

Where mammals exist in herds or flocks, recognition of the species is by sight or smell, or more usually by both simultaneously. We humans, who are strongly visually oriented, are apt to think that other mammals use their sense of sight more than they do. Actually, most mammals are more apt to identify their fellows by odor than by body pattern. To be recognized as a member of the same species by most mammals, it is necessary that an animal smell right. A wag once defined a species, in canine terms: "A dog is an animal recognized as a dog by another dog." A student of animal communication would modify this to: "A dog is an animal that smells like a dog to another dog." Certainly, anyone who has a dog knows the importance of odor recognition to this particular mammal.

Social aggregations

Specific aggregations have evolved into elaborate societies in the social insects. An ant or bee colony is composed of large numbers of individuals living together in a nest or hive. Usually there is one reproductive individual, the queen, who is the mother of the whole group, and large numbers of permanently immature female workers, who do the work of the colony. A few males may be kept around, or produced at appropriate times, to fertilize the queen.

Worker ants and bees hunt for food in all directions from the colony. Upon their return, they must be recognized. For this, they have a distinctive species and colony odor. Individuals of other species or from another colony are summarily driven off or killed. Individual worker ants have the colony odor which is acquired by the ant in mingling and feeding with others. Some species of ants capture ants of other species and hold them as slaves; the "masters" and slaves quickly acquire the same odor, and form an harmonious colony.

Bees also possess hive and individual odors. The hive odor acts like a password that enables a bee to get into the hive and not be challenged or killed. Recent work has indicated that the hive odor has a minimum of twelve different chemical components, and is affected, at least in part, by the food that the bees eat. It is passed from one bee to another when the insects exchange food. It is not inherited, for it is quickly acquired by a foreign bee, if the foreigner is introduced into the hive in a cage to protect it from damage by the other bees. As soon as the strange bee acquires the hive odor from being near the others, it can be released from the cage, and the others accept it as one of the group. The sense of smell in bees is remarkably acute, and the hive odor can be detected at incredible dilutions. If a worker bee is allowed to walk around in a glass tube to mark the tube with the hive odor, its hive-mates can recognize the tube weeks afterward, when it would seem that just about none of the scent should remain. To man, at least, it has long since disappeared.

Interspecific aggregations

In our definition of communication, we stated that we would, in general, restrict the term to transmission of information within a species. There are, however, a few cases in which members of two species live together, as if they were one, and in these cases there are methods of communication that fit the situation.

For instance, some coral reef fish live among the tentacles of large sea anemones. Generally, the sting of sea anemones kills fish, but, in this case, the sea anemones "recognize" the fish and do not sting them; so the two live amicably together. In the same coral reefs, gaudily colored, banded shrimp, called cleaner shrimp, feed by picking parasites off the fish. The cleaner shrimp stand on the tips of the coral heads and wave their brilliantly colored antennae and feet back and forth; the fish, attracted by this signal, line up to have their parasites removed. The shrimp are quite palatable to the fish, as can be shown by crushing them and feeding the meat to the fish, but the fish actually allow the shrimp to walk unmolested in their mouths to clean their teeth. If danger comes, the fish eject the shrimp before swimming away.

One might wonder whether any species of animal imitates the shape, smell, or sound of another species to attract it for food—a literal "wolf in sheep's clothing." We know of no case of this, and probably for a good reason. If a race of cats, for instance, evolved so as to become more and more like mice, until finally they smelled like mice, other cats would undoubtedly prey upon them, just as if they were mice. A predator that would adopt the exact odor, sounds, or pattern of its prey would, by that ruse, expose itself to predation by its own relatives.

A situation akin to disguise, however, occurs in some beetles which live in ant nests. The beetles accept the hospitality of the ants, without adding anything; they are unbidden, useless guests. Most of these beetles are antlike in shape, and covered with a tough cuticle. When they first enter the colony, the ants pull and push them about, and try to bite them, but their armor-plate protects them. The beetles merely pull themselves into compact masses and wait. When they have been in the ant nest for a few days, they have acquired the colony odor, and so give the correct password to the ants; they are fully accepted after that as members of the group.

DISPERSAL SYSTEMS

Although many species of animals live in aggregations, many others disperse themselves, so that food and space are more or less evenly distributed. Unless this results from actual food shortage, or from fighting, it too requires means for specific identification.

In many lower animals, spacing out occurs directly, through scarcity of food. For example, if too many marine worms settle at one place on the ocean floor, they use up food so rapidly that some starve. The result is a gradual spacing out. This does not require any particular method of communication, but it is, at least to man's way of thinking, rather wasteful and haphazard.

In other animals, such as many fish, spacing out is achieved by fighting, but this too often results in loss of life.

In many groups of animals, behavior patterns have evolved that allow the animals to maintain spacing without actual damage. True fighting gradually becomes what is called ritualized fighting, in which the animals fence with the weapons that they use for fighting, but do not inflict wounds. Further evolution results in aggressive displays, in which the animals either display their armament, or indicate their aggressive intent by threat postures or sounds. They do not come to blows, and thus there is no danger to the individuals. Finally, spacing is enforced by the development of territoriality. In this case, an animal produces some signal —pose, sound, or scent—to announce that its place of residence will be defended; generally other members of the species, so warned, do not trespass on its territory.

Ritualized fighting

Ritualized fighting is really not an example of communication, unless we consider it tactile signaling. Generally, the two animals

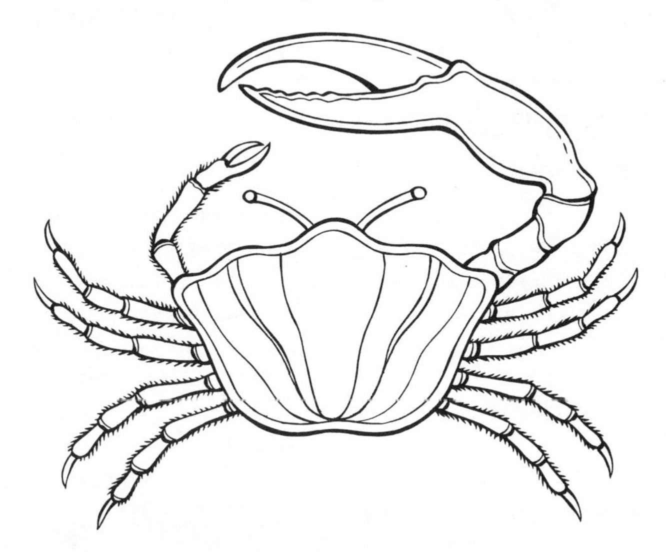

FIGURE 6. *Male fiddler crab, showing enlarged claw (fiddle). About natural size.*

indulge in a fair replica of ordinary combat, but they do not seriously use the jaws or claws which they have.

A number of species of crabs, fiddler crabs (Figure 6), for example, engage in ritualized fighting to contest for specific places. They seize each other by the claws as if with murderous intent, but do not actually inflict wounds. Many other crabs do not have ritualized fighting, and actually tear claws or legs off each other in serious fights. Even within the same species, some individuals seem to be much more given to real fighting to maintain their rights than do others. If a nonaggressive individual retires soon enough, the fight has the appearance of a ritualized fight, but if the intruder does not retire, the fight is quite in earnest.

Ritualized fighting is seen at its best in birds and mammals. As an example, "Gooney Birds"—or more correctly, Laysan Albatrosses—which make their nests on mid-Pacific islands, show highly developed territorial behavior. The occupant of a nest, if approached by an intruder, rises and, after a series of communication signals to notify the intruder of its trespass, lunges at it. The two next engage in beak-fencing, with considerable vocalization. Generally, at this point, the intruder retires without further pressing the issue. Among many mammals that could be cited, pigs and goats push or butt each other, without actual damage being done to the contestants, to establish an area of dominance.

Aggressive displays

Aggressive displays, which do involve communication, are usually evoked by some immediate stimulation to the displayer—usually the presence of an invader in a defended region. This distinguishes them from territorial behavior, in which the invader need not actually be present for the display to occur. Generally, aggressive displays are visual, but they may have auditory components as well.

Although the claws of crabs and crayfish look quite formidable, and can be used for actual fighting, they are mostly used for aggressive displays. Many crabs have oversized and brightly colored claws—real display pieces. When a male fiddler crab is approached by another male or female, it waves its large claw in a characteristic motion that identifies it clearly to the intruder. This is an aggressive signal, a warning to the intruder that it has come into a defended region. If the intruder is a smaller male, or a female that is not sexually receptive, it usually retreats. If it is a larger, more aggressive male, he answers the display with similar motions, and, unless one retreats, there is a mock or real battle.

The spiny lobster has small claws which could not be used in displays. It has, however, at the base of each antenna, a comblike ridge which, when rubbed against the carapace, produces a grating noise. This noise acts as a defensive signal when another spiny lobster invades its home area, identifies the species, and informs the intruder of its trespass.

Many terrestrial arthropods that maintain areas in which they forage also show aggressive displays if the areas are invaded. Some beetles have large jaws which they spread menacingly as a threat to intruders. Honey bees are well-equipped for actual combat, as everyone knows, but they also have an aggressive display—loud buzzing—if intruders try to enter the hive.

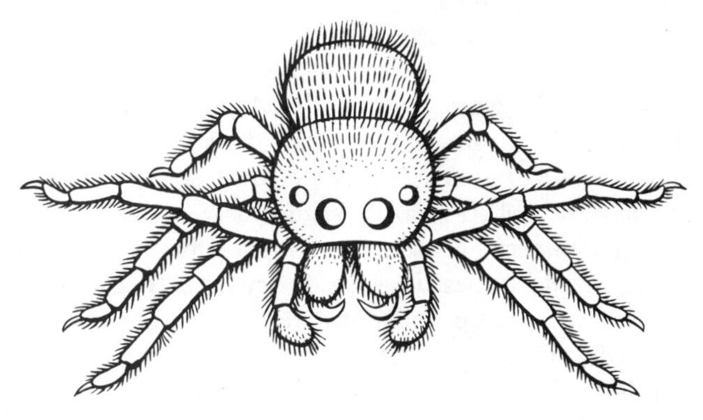

FIGURE 7. *Male jumping spider, displaying front legs. About 6X natural size.*

Among the most elaborate of aggressive displays are those of male jumping spiders. The jumping spiders do not spin any permanent web, but wander about hunting prey. If two males meet, they threaten each other with their hairy front legs and large eyes.

The displays are entirely visual (Figure 7), the spiders brandishing at their opponents brightly colored hairy patches on their legs. Some species also alter the position of the pigment in their eyes, so that the eyes flash in garish colors at the intruder. Generally, the displays are sufficient to cause the males to separate without fighting, but occasionally a fight ensues. Usually it is ritualized fighting, and no one is killed, although the contestants carry poison in their fangs.

Many fish guard regions in which they feed or have their nests. Male European sticklebacks, for instance, defend areas in aquariums. If another male approaches, the defending male adopts a vertical stance, as if standing on his head. If the second individual continues to approach, the first attacks. Usually, however, the second retreats from the threat display. The popular aquarium fishes, swordtails and siamese fighting fish, show excellent aggressive displays when two males are put near each other. Even if the males are separated by a glass plate, they flash their colors and make aggressive passes at each other. If the two fish are not separated and are about evenly matched, the aggressive displays soon become a serious fight, which can result in death.

Aggressive displays are quite common among snakes and lizards too. The erected cobra, with its hood spread, is engaging in a visual threat to protect its territory. Usually the message gets across, not only to members of its own species, but to other species as well. Other snakes writhe in what have been called dances—sometimes mistaken for courtship behavior—when their areas are invaded. Lizards, such as iguanas, thrash the tail, or spread their bodies up and down to look bigger than they are.

Hundreds of examples of aggressive displays could be cited among the birds and mammals. Gulls and albatrosses, for instance, when approached by another member of the species, pull themselves up as tall as possible, the upright threat posture, which signals to the approaching bird that it is entering a defended area. The threatening stance may also be accompanied by bill snapping, which is probably a notification of sanguinary intent. If the intruder continues to approach, the aggressive display changes to ritualized fighting, or sometimes to actual fighting. In threat displays, the birds indicate clearly the intensity of their aggressive intent by changing patterns. A bird can erect itself only partly and snap the beak quietly, or it can draw itself up to full height and scream defiance.

A similar situation exists in mammals, in which aggressive displays usually are intention movements, that is, are derived from actual combat patterns. The baring and snapping of the teeth, growling, and bristling of the fur are characteristic of mammalian aggressive displays. In male moose and elk, there are characteristic stiff-legged gaits which indicate to other males that they have entered a defended territory. Special gaits or stances in mammals are often accentuated by growling or shrieking, all acting to intimidate the intruder.

Territorial behavior

The most highly developed means for maintaining spacing between individuals of a species is the development of territorial behavior. In this case, the animals produce signals, without actually being in sight or contact with another member of the species, to notify others of their rights.

One example would be the songs of crickets and grasshoppers. The ordinary song of a male cricket, although used also to call in the female, indicates a territory that the male will defend. If another male approaches, the singing male, on catching sight of the intruder, changes from the calling song to another, which has been named the rivals' song, an aggressive display. The two rivals then sing at each other with this song, and the matter may be decided then and there, with one of them moving off. If, however, the two prove equally persistent, they engage in ritualized fighting, or even serious fighting, resulting in injury or death. In most encounters, the relative intensities of the rivals' songs apparently sctttlc thc issuc without resort to combat.

The best studied of all types of territorial behavior is that of birds, and it is here that the concept of territoriality was first developed. The song of a male bird is an announcement of territory, not, as many people believe, a sign of pleasure. An American Robin, for instance, returning to his northern summer home after winter in the south, selects a territory where he and his future mate can have a nest. This he is willing to defend with actual fighting, if necessary. Usually it is quite unnecessary, for his song announces to all other males in the neighborhood that he is defending that particular spot. The male Robin need not see other individuals to sing persistently.

A male bird's song is not only an announcement of territory to other males, but is also an announcement of his mating intention toward females. In this regard, it is exactly like the cricket song in its informational content. If another male bird approaches, the singing male changes from the territorial song to an aggressive display which usually involves an upright threat posture, threatening calls, and passes at the individual. If this does not result in the departure of the intruder, the male resorts to ritualized or real fighting. Generally, however, the territorial song maintains distances between individuals and space for breeding, without biologically disadvantageous fighting. In this way, more individuals are spared for reproduction.

If one broadcasts a recording of his territorial song to the same bird in his own territory, he comes to the loud-speaker and displays great vocal talents, as if to outdo the recording of his own voice. However, lacking a visual dummy, the male bird does not proceed to aggressive displays or fighting.

Many mammals also maintain regions in which they are dominant by the use of communication signals. Auditory signals are common. Bull seals, for example, stake out territories in regions where they expect to breed, and show their intention to defend them by loud roars. Similarly, squirrels chatter and make churring noises to indicate their territories. Primates, particularly gorillas and apes, vocalize, or pound their fists on the ground or against their chests, to announce their intention to defend a given region.

In line with the strong olfactory bias of most mammals, chemical signals are also used to stake out territories. Chamois and ground squirrels, for instance, deposit odors as they move about, and these mark the areas as out-of-bounds for other individuals of the species. Anyone who has walked a dog on a leash is familiar with the dog's use of its excretion to produce an odor marking its particular territory, as well as to transmit other information, such as about its sexual state.

A territory is usually maintained inviolate only within the species, and, therefore, species recognition is quite necessary. Unless animals of different species compete directly for the same items of food at the same time they usually ignore each other. A number of birds may live in a small tree, but generally not two members of the same species. If, however, animals of different species compete at the same place and time for the same item of food, there may be contention and the use of characteristic aggressive displays. Where different species feed together, they may learn each other's signals, and interspecific communication may develop. Ordinarily, however, size and force determine the outcome.

An interesting situation, in which one species invades the territory of another by utilizing the correct communication signal, involves the Cowbird. This bird lays its eggs in other birds' nests and leaves the other birds to raise its young. Ordinarily the appearance of a strange bird at a nest would induce aggressive reactions. The Cowbird, however, turns aggressive displays aside by assuming a submissive posture, soliciting preening. Many birds have these special positions to induce their partners to tidy their feathers. The Cowbird, by assuming this totally submissive position, as soon as an aggressive display occurs, turns aside the aggression; it is thus able to invade the other bird's territory without incurring a fight. It is interesting to note, in this regard, that the Cowbird, like the European Cuckoo, which has the same "parasitic" habit, is drab and rather nondescript in color. It thus avoids bearing any bright identifying colors, which might release specific aggressive actions from its prospective victims.

Animals generally live in complex biological environments, with members of other species as well as their own pressing in on all sides for space and food. Even when they are not reproducing, animals need to be able to identify members of their own species from others, for their own kind represents the keenest competition for the necessities of life. Whether the ultimate result is social cooperation between individuals, or peaceful separation, animals must be able to identify their fellows. This they do through the use of signals, sign stimuli, transmitted through the sensory channels best adapted for the purpose—specific movements, specific odors, specific sounds, improbable enough in the environment to mark the producer.

5

Social Cooperation

FEW animals exist entirely independently of other members of the same species. Some exist in aggregations of variable structure, whereas others maintain, at least part of the time, a certain degree of distance between the individuals. In either case, it is necessary that they be able to identify themselves to others of the species, as we have seen.

Regardless of the degree of organization, however, many social relations and interactions occur, other than mere maintenance of cohesion or distance. Two important items of information that many animals, even if nonsocial, transmit to other members of the species are warning of danger and notification of food sources. Signals that aid in these relationships are important in social cooperation.

ALARM SIGNALS

It is difficult for human beings to realize the world of danger in which the average animal lives. All around are enemies which would kill it for food or compete with it for space. It is, therefore, of great biological value to have some means whereby an alarm can be spread, when a prospective enemy appears, and many species have alarm signals.

Many animals living in the depths of the ocean, where there is no light, produce luminous signals, using complicated light-producing organs. There has been considerable debate about how these signals are used. Some biologists believe that they are important in mating. However, the signals are often produced by animals while not in mating condition, and, therefore, would seem to have other functions. Among these, it has been suggested that they act as burglar-alarm systems, that is, when the animals are attacked by predators, they flash, and thus warn others of impending danger. Proof of this, however, must await future opportunities to observe deep-sea life.

The most complicated alarm systems are found among the social insects, ants and bees. Among ants, and some termites as well, notification of danger to the nest seems to be transmitted by three different routes: tactilely by wild alarm dances, chemically by production of alarm scents, and auditorily by production of rasping sounds or tapping of the body against the side of the nest. Many species of ants and termites have specialized stridulatory ridges on the abdomen, and they can produce high pitched squeaks by rubbing the abdomen against the thorax. In some species, at least, this is done only when the ants are disturbed, and may spread an alarm through the colony. Ants do not seem to have receptors for sounds, in the usual sense of the word, but they have very sensitive vibration receptors in their legs. It is quite possible, therefore, that the sense-organs in their legs receive, through the nest floor, vibrations induced by their stridulation.

Much more is known about scents that spread alarm in ant colonies. These may be produced by glands located near the mouth-parts—the mandibular glands, or by glands in the abdomen—the anal glands. In the ant, *Tapinoma nigerrimum*, the actual materials that are produced to alarm the ants have been chemically purified and identified. These are two organic compounds: methylheptenone and propyl-isobutyl-ketone. Pure samples of these chemicals cause the same alarm reactions, when exposed near ants' nests, as do the substances secreted by the ants themselves.

It has been suggested that chemical substances such as these, which are used as signals between individuals, be given a special name, pheromones. It is believed that these substances act somewhat as hormones act in an individual animal's body—they are chemical materials given off by one unit of a living organism, and carried in the environment elsewhere to produce an excitation. Some species of ants, however, produce the same results without using odors, by running about or producing exciting sounds. The name, pheromone, would seem, therefore, to be somewhat specialized, for chemical signals are merely one class of communication signals used by animals.

Honey bees, similarly, spread alarm through the colony by agitated running about, and by an odor, that of bee-stings. It has recently been shown that one of the chemicals in bee-stings which excites honey bees is iso-amyl-acetate. This is a rather common paint solvent, banana oil. Apparently, however, this is not the only element in the odor of the bee-sting that is effective, for, although exposure of this near the hive excites the bees, it does not induce stinging; exposure of a used bee-sting does both. As in the ants, the multiple nature of the alarm process in bees is indicated by the fact that there are also sounds produced by excited bees. Bees do not seem to be sensitive to air-borne sounds, but they are extremely sensitive to vibrations, and probably receive the alarm signals as vibrations in the hive.

Chemical alarm signals are widely distributed in the Animal Kingdom. Fish, for instance, when injured give off an alarm material which causes other members of the species to swim away and hide. This material from the skin, called *Schreckstoff* by its discoverer, the great German zoologist, Karl v. Frisch, is extremely potent. As little as 0.002 mg. of fish skin (about 0.01 sq. mm.), in an aquarium holding 14 liters, can alarm a group of minnows. The material is not exactly species-specific, for it alarms fish of other species as well. However, the degree of alarm seems to be related to the closeness of relationship of the fish. Interestingly enough, it has been shown that the alarm chemicals given off by injured or disturbed marine fish are attractive to sharks. Thus, the alarm signal, while possibly saving the group, can result in the death of the signaler.

In line with their general tendency toward the use of sounds, birds spread alarm mostly by calls. Some species have a number of calls for the purpose. For instance, small European finches have two alarm calls. One, the so-called hawk call, is given when they sight a hawk overhead. It is of high frequency and delicate timbre, starting and stopping gradually. The other, the ground call, is given when they see danger on the ground. It is loud, sharp, and staccato. The hawk calls of many species of birds are similarly of high frequency and rather gradual onset and stop. This makes them difficult for a prospective predator to localize, and thus minimizes danger of capture for the alarm giver. Among the few birds that lack alarm calls are the larger sea birds, such as albatrosses. These, however, have always lived over the seas or on deserted, waterless oceanic islands, where they have no effective predators.

Birds react to alarm calls either by flight or immobility, depending upon species, age, and situation. Some birds, the United States Eastern Crow is one, have assembly calls or mobbing calls, which they use when they sight a predator, such as an owl or a cat. These calls attract large numbers of the birds, all shrieking at the enemy. Obviously, there is safety in numbers, and the mob seems to confuse the prospective predator, making it incapable of doing any particular damage. This is, again, like a burglar alarm system, in which the enemy is brought to light and kept in sight.

Many birds that live in flocks have visual alarm signals. These are usually flashing displays of tail or wing feathers. Birds, such as doves, quail, or mynahs, that feed in groups on the ground, have bright outer tail feathers or bright wing patches which are displayed vividly when they fly up in alarm. Even birds that seem, on casual observation, to use auditory signals almost exclusively may also use visual signals. European Bullfinches, for instance, have alarm calls, but they can still transmit alarm from one to another when totally deaf. Obviously, besides the auditory signals, they must have visual signals as well.

Mammals have a wide variety of alarm signals, utilizing all their well-developed senses. Many species of deer, for example, have a characteristic white patch beneath the tail, which they flash as they run from approaching danger. Other deer have alarm odors, which they give off to alert their fellows.

As in the case of birds, auditory displays are also used by mammals to spread alarm, and they are particularly useful, because they carry for great distances. Many examples could be cited. Moose, elk, and deer emit a strong bark when moving away from a danger area. They also use, as a visual signal, a gait adopted only when they are alarmed. The European chamois has a characteristic alarm whistle, and many primates, such as monkeys and baboons, bark or shriek in alarm. The giraffe is usually considered to be voiceless, but it announces the presence of danger by a sharp snort. Giraffes also, as do many other mammals forming herds, spread alarm by wildly stampeding; this could be considered as visual or tactile communication.

Both birds and mammals, in alarm signals, take advantage of potential variability in intensities, particularly in auditory signals, to indicate low or high levels of alarm. The Herring Gull's use of its alarm call, a series of stacatto notes, illustrates the point. If the birds are approached by an enemy while they are relatively safe on the water, they give the alarm call at low intensity, sounding as if they are just clearing their throats. If, however, they are frightened while on land, as when a human being suddenly approaches them, the alarm notes are loud, piercing, and sharply staccato. The reactions of their fellows are directly related; to low intensity alarm notes Herring Gulls merely become attentive and look around; to high intensity notes they fly up immediately, circle around, and then depart. Similarly, with mammalian signals, such as the alarm calls of deer: these may be low, alerting barks, or sharp, warning yips, depending upon the degree of danger. Here, too, the reactions of the receivers are related to the intensities of the signals.

DEPARTING SIGNALS

Some time ago, the German biologist, Faber, reported an interesting variant on the alarm signal. He noted that, whenever he moved about in a field, all the grasshoppers jumped at his approach. Yet, the grasshoppers themselves jumped about from place to place all the time without alarming each other. Close listening revealed the secret. Before jumping, each grasshopper gives a short set of clicks, called by Faber the departing song, which conveys to the others the information that it is going to jump.

Since that time, many species of grasshoppers have been found to have departing calls. If they jump without giving the departing call, as they do when a human being or some other enemy approaches, all the others jump with them. The rustle of the jumping individual, when it occurs without prior notification that it is going to happen, alarms the grasshoppers.

It seems probable that some birds and mammals have similar signals, although studies of this are almost nonexistent. Under normal conditions, meadow mice are said to squeak at frequencies which are ultrasonic for man, and therefore inaudible to him. However, if a prospective danger arises, they run noisily, without making the squeaking sound, and other meadow mice nearby are alarmed. Eastern Crows, when feeding in a tree, come and go without alarming the whole group. Before leaving, if no danger is present, they emit a short set of notes, which apparently act like the departing song of grasshoppers. If a Crow suddenly flies away from the group without making any sound, the others usually take off with it. It has silently spread an alarm—the fastest and safest way. One can read between the lines so to speak, in published descriptions of behavior of other birds, and find indications of departing songs. Queleas and Weaver Birds in Africa, and

Mynahs in Asia, for instance, produce faint calls before flying away from a group, unless they are alarmed.

DISTRESS SIGNALS

Alarm calls are usually emitted by individuals that sense danger, but are not actually being injured. Many animals, however, also emit calls when they are actually being attacked. These distress calls might possibly cause the predators to release their prey; they also can warn other members of the species.

Some insects have been said to make special sounds or exhibit special behavior when they are injured, but little is known about this so far. Some species of ants and beetles, when held, use their stridulating ridges to produce squeaking notes. These may scare their enemies, but, at least in ants, they also may draw other members of the colony to the calling individual. Cicadas, which sing lustily in trees in the summer, buzz stridently at birds that chase or seize them. The buzzing note may cause the birds to release the intended victim. Many insects, when roughly handled, emit chemicals that act as repellents. These, however, are not true communication signals, because they irritate or harm the attacker. Chemical analyses of these repellent compounds has shown them to be harmful substances, such as benzoquinones or benzaldehyde, even hydrocyanic acid gas.

The use of distress signals for communication apparently is common in birds. As with alarm signals, they are usually auditory. Almost all birds have distress calls, generally raucous shrieks, which they emit when handled roughly. The reactions of birds to the distress call of the species are related to social organization. In species whose members live mostly out of contact with each other, where the distress of one individual does not mean a real threat to another, or in species in which there are no effective predators, such as the giant Albatrosses, there are generally no reactions at all to distress calls. In other species, where individuals live moderately dispersed, such as Jackdaws in Europe, or Crows and Robins in the United States, the distress call attracts other members of the species, bringing about mobbing of the prospective predator. In birds that live in compact flocks, where distress to one means potential danger to all, the distress call is strongly repellent. In Starlings, for example, which aggregate in large flocks, where a predator could easily turn from its current victim to another, the reaction to the distress call is dramatic. Whole flocks of birds can be turned away by simply broadcasting to them a recording of their distress call.

Mammals also favor the auditory channel for distress signals. Margaret Altman reports the case of an elk calf which had fallen into a ditch in the Jackson Hole Wildlife Preserve, and set up its distress call. This spread excitement to all the elk around, and they came to the ditch to gather around the distressed youngster. Baboons and other primates are attracted to screaming members of their species; screaming is the distress call. As a matter of fact, some investigators have noted that a man may escape attack by one of the larger primates by screaming. Chimpanzees have been known to turn from aggressive behavior toward a human being to one of solicitation or rescue, if the person screams, as if in distress. Rats and mice, which live in large, loosely organized groups, even though individuals generally remain apart except for breeding, seem to have distress calls which are ultrasonic for man and so cannot be heard by him. These rodents hear at least two octaves higher than man, and therefore can hear these calls. They respond to them by scurrying for their holes. It has been reported that even dolphins and whales have calls that seem to be distress signals.

Distress calls of the higher vertebrates sound much alike. Even an untrained person listening to a recording of the distress call of a bird or mammal usually identifies it as a distressful sound. Biologically, of course, this interspecific overlap could have considerable value, for danger to any species is likely to be danger to others of approximately the same size. An interesting case from our own laboratory illustrates this. We had recorded the distress call of a cottontail rabbit, to see whether it might attract their predators, foxes. However, when this was broadcast, it attracted mostly owls, and many of these were too small to have attacked a rabbit, even an injured one. It is quite likely that the owls responded to some general distress significance, and that they are thus usually attracted to potential food.

WARNING SIGNALS

Warning signals of animals are produced by species that are poisonous or distasteful, and serve to warn prospective enemies. They might be considered as interspecific signals, usually involving learning by the prospective attacker that the warning species is dangerous.

Venomous animals, such as snakes, scorpions, wasps, or poisonous caterpillars, are often either brightly colored, or able to produce striking displays. The buzzing of rattlesnakes, the hood-spreading of cobras, and the posturing of scorpions are good examples. The skunk, which is certainly well-armed, is strikingly colored, and uses its flicking tail to warn an intended target. Many primates, perhaps even man, have an inborn dread of snakes, even snakes that have no characteristic patterns or displays. Web-spinning spiders usually release wasps when the wasps start their warning buzz; in this case, the spiders cannot have learned through experience that this represents danger, for the wasps can kill the spiders summarily.

This sort of signaling protects the warning animal, as well as the animal that is warned. In most cases, the warning display is either such as to be alarming in itself—the rattle of rattlesnakes, or the wicked buzz of wasps, for instance—or the individual being warned has learned to avoid this particular pattern because of previous experience with animals colored like this. As an example, young birds at first try to feed on brightly colored caterpillars, which are distasteful or have poisonous spines, but, after a few disagreeable experiences, learn that gaudy caterpillars are dangerous and avoid them.

The warning coloration of dangerous animals is exploited by some nondangerous animals for their own protection; this is called mimicry. Monarch butterflies are distasteful to birds and are prominently colored. Generally birds that have had any experience with them avoid butterflies with the same color pattern. Viceroy butterflies, which are edible, are patterned almost exactly like monarch butterflies, and apparently, by this means, escape being eaten. Monkeys that try to eat distasteful or noxious caterpillars with striking patterns do not later touch caterpillars with similar patterns even though these are harmless.

These bluffing signals may become complicated. There are certain flies that are colored and shaped like venomous wasps. Not only do these flies look like wasps, they also buzz like wasps when molested, a form of audio-mimicry. Some of these flies go still further, driving the abdomen down on an enemy, as a wasp would do in stinging, even though they themselves have no sting. All this complex of signals, when flashed on an animal that has had previous experience with wasps, is apt to cause the animal to release the fly. A harmless snake, *Dasypeltis*, likewise, rolls itself into a formidable coil and then, by moving special scales of its body against each other, hisses menacingly, thus acoustically mimicking dangerous species.

FOOD SIGNALS

Many animals share information about food sources with others of their own species. Naturally most animals do not, for food is a necessity of life, and an animal tries to keep what it has for itself. Signals indicating food sources are generally used by social animals, but this is not exclusively the case.

Flies are not usually considered to be social, for they are not organized into colonies. Yet flies mark foods with a special odor, called the fly factor, which is attractive to other flies. If one exposes a piece of sugar near flies, it may be some time before one finds it; as soon, however, as one does find the sugar, it marks the food with its odor, and shortly the sugar is covered with flies.

This situation is similar to that in the Herring Gull, Chukar Partridge, and gibbon. These animals, among others, have special calls—food-finding calls—which announce food sources to other members of the species. Herring Gulls and Buzzards also signal food locations by special flight patterns above them.

GUIDANCE SIGNALS OF ANTS

From these rather simple notifications of food sources among nonsocial animals we may turn to the elaborate systems of communication about food of the social animals, such as ants. The colonies of ants, even though they number thousands of individuals, depend upon a few scouts to find food. The scouts then recruit the workers in the colony, all permanently immature females like themselves, to collect the food.

A scout ant often finds food at some distance from the nest, and so must not only proclaim her find, but must also indicate where the food is. Generally, the scout discloses the discovery by agitatedly running around, waving her antennae. The agitation gradually spreads, and the ants gather around and stroke the excited scout. In many species, the scout now leads the alerted ants to the food, a very simple type of communication indeed. In other species, however, the scout, during her trip back to the hive, has deposited a series of odorous droplets to form an odor-trail. This is like dropping pebbles along a woodland trail, so that one can later find his way back. The recruits follow the trail, and thus reach the food, although the scout remains in the nest. As more and more ants join the procession to and from the food source, they add to the odor-trail and it becomes more and more distinct.

There has been much debate among students of ant behavior as to whether the odor-trails are polarized or not, that is whether they indicate direction. A number of experiments seem to indicate that the trails are polarized. For instance, if ants make a trail crossing a turntable, and the turntable is then turned through 180°, so that the trail is reversed, the ants, on coming to the edge of the turntable, stop and are disoriented. Obviously, if the trail were like railroad tracks—could be used equally well in either direction—then the ants would use it either way. This, however, is not the case. This indicates that there is some degree of directionality about the trail, but just what it is has not been determined.

A human being finds it difficult to understand how a droplet of scent could indicate direction. However, if the drop of scent had a special orientation or shape, it might perhaps be like an odorous arrow. The great student of ant behavior, Forel, suggested that ants have a special chemical sense, which he called the topo-chemical sense, allowing them to distinguish not only the odor, but also simultaneously the shape of the drop. We are quite familiar with this in human vision, in which we distinguish, for instance, a yellow square. Forel pointed out that we have two separated eyes which allow us to distinguish the shape. Similarly, the ant has two separated antennae, with which it smells, so it too might be able to distinguish shape chemically. An ant might actually smell, for instance, an ant-odored ellipse, in the same way as we distinguish a yellow square. It could thus determine the topography of an odor-trail and orient to it. This is an interesting idea, but has never been proved.

It is known that ants use a light-compass reaction to guide them in returning to the nest, so that the odor-trail could be used to show the path, and the sun to show direction. In the case of the turntable experiment, it might be that the two ends of the odor-trail were not matched. A study of this type of orientation in ants, using modern techniques, should yield some interesting results.

The use of odor-trails as guides to food sources might seem to be disadvantageous, in that odor-trails are persistent, and therefore could lead the ants to exhausted food areas. Actually, except under very damp conditions, the odors generally evaporate rapidly. However, even where the trails remain for some time, they may have value in leading ants back to former food sources which may again become rich, or in leading wandering species, such as army ants, back along their trails, or toward other colonies.

GUIDANCE SIGNALS OF HONEY BEES

The most exact signals used as guides to food sources have been developed by honey bees. Their system is so elaborate and complicated, that its discoverer, Karl v. Frisch, has declared that a scientist is duty-bound not to believe it, until he is forced to do so. It now appears, as a result of the brilliant studies of v. Frisch and his students, that we are driven to believe in the remarkable system of communication that honey bees have developed.

Guidance to sources of nectar

What information would a scout bee that has found a rich source of nectar at, let us say, two miles from the hive have to give her hive-mates, if they were to find this efficiently? Obviously, the scout cannot merely indicate that she has found the food and give no other information. It would take too long for the other bees to scour an area with a two-mile radius trying to find the food.

The scout gives the information that she has found something by a series of rapid movements, bumping into the other bees. What she has found is communicated by odors on the scout, for she has picked up the scent of the nectar-bearing flowers on her body. These two items of information are good, but they are not good enough. If the bee colony is to work efficiently, the other bees must be informed of the distance and direction to the food source.

The scout bee could lead the others to the place, as scout ants do, but it is difficult for flying bees to keep any sort of organization in a column. Thus the bees have developed means for the scouts to communicate to others in the hive the direction and distance to the food. A further item of information, which, although not absolutely necessary, is very valuable, considering that there are many scout bees out at one time, has to do with the richness of the food source. Therefore, if three or four scouts arrive simultaneously with different food source discoveries, their hive-mates could determine which one of these was the best to exploit.

All of these items of information are transmitted very exactly by honey bees. The nature of the nectar source is shown, as we noted, chiefly by the odor of the flowers on the body of the scout bee. Where flowers do not have odors, or where, experimentally, dishes of sugar water are used, the scout bee marks these with a scent produced by special scent-glands on her body. This is similar to the use of the fly factor by flies. One of the components of the bee's scent is geraniol, but this is not the only odorous material secreted by the glands.

The distance to the food source is communicated by special movements, which v. Frisch has called dances. In the German race of honey bees, two types of dance are used. If the food source is no more than 275 feet away, the scout bee performs what is called a round dance, going round in circles alternately to the right and to the left. The other bees follow the scout, and are excited to leave the hive and fly around searching for the flower designated by the odor on the scout's body. If the distance to the food source is greater than 275 feet, the scout performs another type of dance, called the waggle dance, or wagtail dance. In this, she runs in a straight line for a short distance, wagging the abdomen vigorously, then makes a semicircle, waggles in a straight line again, makes another semicircle to the other side, and turns to make the straight waggling dance again. This too the hive mates follow, and from it get the information they need.

The tempo of the wagging dance indicates the distance. If the food source is about 1000 feet from the hive, for instance, the bee makes thirty complete runs per minute. If the food source is over 2000 feet from the hive, the bee makes only twenty-two complete runs per minute. Actually, four items in the dance vary with the distance: the time for the complete circuit, the duration of each waggle of the abdomen, the length of the wagging run, and the number of waggles per circuit. The most exact measure, as far as has been determined, is the duration of the waggle, with the time for complete circuit almost as good. It has recently been found that the dancing bees also produce a sound of about 250 cycles per second during the waggle dance, but what information this transmits is unknown. The hive-mates of the scout follow the dancing bee and feel her with their antennae, for it is perfectly dark in the hive and they cannot see her. In this way, they receive the information about the distance to the food.

The waggle dance, furthermore, indicates the direction to the food. If the bees, to get to the food, should fly directly toward the sun, the scout makes the waggle run vertically upward on the honeycomb. If they should fly directly away from the sun, the scout waggles vertically downward. All angles between these are indicated by equivalent angles from the vertical on the comb. The honey bee thus transposes a horizontal direction, based on recognition of position of the sun, into a vertical direction, based on recognition of gravity.

In some cases, honey bees perform their guidance dances right at the entrance to the hive, where the sun is visible. In this case, they point the waggling run directly toward the food source. The bees do not have to be able to see the sun itself when dancing on a horizontal surface to be able to indicate direction; they need only see part of the sky. Actually, they do not use the sun itself as a compass, they use the polarized light pattern of the sky. The compound eye of the honey bee is made up of hundreds of simple eyes. Each of these is essentially a little polarizing analyzer. The scout, by the angle on the comb, transmits to the other bees the patterns of light and dark that should be created by the analyzers in the simple eyes, if the bee is facing in the right direction. Since v. Frisch's discovery of this property of the honey bees' eyes, many insects, and other invertebrates also, have been found to be sensitive to polarized light. The human eye, however, is not sensitive to the plane of polarized light, and so man finds this type of reception difficult to understand.

Obviously, if the scout bee is to indicate the direction to the food, and if she is to continue dancing for some time in the dark hive, she must have some sort of built-in clock, so that she can change the direction of her dance as the sun changes its direction in the sky. This she has. If the bee continues to dance for two hours, she changes the angle of the waggle run such that, at the end of the time, it is still giving the correct direction to the food. She does this, even though she cannot see the sky or sun.

The time sense of honey bees is most remarkable. One of v. Frisch's students, Lindauer, trained honey bees to go to a certain source of food. They were then brought inside a building, because of cold weather. After the hive had been indoors for thirty-eight days, food was given to the bees, whereupon the scout bees signaled the direction to the old food source, even though they could not get outside. The amazing thing was that if they were fed at different times of the day, they gave the correct sun angle for every time of the day, although they had not seen the sun in their indoor room for all that time. Obviously, they had kept track of the passage of time. Furthermore, if honey bees arc transported from the northern to the southern hemisphere, where the sun moves differently, they quickly learn to keep time in this situation.

The scout bee, then, can indicate the nature of a food source and the direction and distance to it, thus enabling hive-mates to find it accurately. One other item of information, as we noted, would be valuable. This is some indication as to the value of the food source. The scout bee transmits this information by the vigor of the dance. If the food source is relatively poor—that is, if the concentration of sugar in the nectar is low, the amount is small, or the distance to the source is great—the bee dances for only a short time and in a rather lackadaisical way. If on the other hand, the food source is rich—that is, the sugar concentration of the nectar is high, the amount is large, or the distance is short—the bee dances with extreme vigor and persistence. Furthermore, as recruits come back from a good area, they take up the driving dance.

If three scout bees, for example, find three different sources of nectar, each at first recruits some individuals to go out. If one, however, has found a poor source, she soon stops, and the returning recruits dance little or not at all. Consequently, this food source is neglected. If one of the three has found a rich source, on the other hand, she continues to dance and dance. The many recruits, when they come back, are also excited and dance vigorously. Gradually, large numbers of the bees are recruited and the hive concentrates on this rich food source.

Guidance to new quarters for swarming bees

Although indication of food sources is of obvious value, the dances of bees play a significant role in another aspect of their lives. Bee colonies reproduce, in a sense, by swarming. A queen takes a large group of workers with her and leaves the old hive to another queen and her workers. The swarm may fly for some distance, and then settles on a limb or some other convenient location. Here the worker bees gather about the queen and remain in a rather compact group. From this, scouts go out to seek prospective home sites. On finding suitable places, the scouts return, and, by dancing on the surface of the swarm, communicate the distance and direction to the prospective nest site, using the same dances as are used to indicate a food source. As in the case of a food supply, their judgment of the value of the find is transmitted by the vigor of the dance. It sometimes happens that two scout bees return with information about home sites that are almost equally good. These bees may then dance for a day or more, before one of them gives up and allows the other one to deliver the final guiding message. During this time, the dancing bees continually change their dancing direction, as the sun moves, to indicate the up-to-the-minute direction to the prospective home site.

Evolution of guidance signals

How such an elaborate system of communication could evolve is a subject of extreme interest. Lindauer has studied the behavior of relatives of the honey bee in an attempt to discover variations of these patterns. He found that bees show a series of communication patterns that correspond with the degree of social organization.

There is one group of stingless bees in Brazil whose colony organizations are simpler than those of honey bees. Some stingless bees, with a low level of colony organization, behave much as ants do. A scout bee, on returning to the nest, simply mills about bumping into the other bees and buzzing peculiarly. The hive-mates pick up the agitation and sound and are excited to leave the nest and seek food. In many such cases, the scout bee leads the group to the food. Then, after they too have found it, they become leaders.

In another group of stingless bees, the scout does not lead her hive-mates; instead, she leaves an odor-trail. She deposits little odorous drops on prominent spots all along the way from the food source to the hive. Then she excites her hive-mates, and they fly out to follow the scent-trail. That this might not be the whole story, however, is suggested by the fact that attempts to make artificial scent trails, using real bee scent, were not successful. The bees easily distinguished these dummies from the real thing.

The scent trail method might seem to be inefficient, but, interestingly enough, it can be used to communicate one item of information that the honey bee communication system cannot. Von Frisch and Lindauer found that honey bees could not communicate up and down by their system. If food for the honey bees is placed on a high tower, the bees come to the base of the tower, but cannot be instructed to fly up to the food. On the other hand, stingless bees that lay an odor trail do not have this problem, for the scout bees simply put scent traces all along the tower right up to the food. Furthermore, the scent trail method is quite appropriate to tropical areas, where unlike the situation in temperate regions, there are few areas clear of trees and underbrush.

The tropical bees must, therefore, go by circuitous paths, rather than by "bee-lines."

From these stingless bees, with odor trails much like ants, Lindauer turned his attention to a rather close relative of the honey bee, the dwarf bee of India, and here he found an intermediate situation. Dwarf bees do not live in a hive, but make honeycombs out in the open. The scout bee needs to see the sun to indicate direction, and she dances only on a horizontal surface. The direction to the food is indicated by a waggling run pointed in the direction of the food on the horizontal comb. If the surface on which the scout runs is turned to a vertical position, the bee cannot "talk"; it cannot translate a horizontal direction into a vertical one.

The communication system of bees, therefore, probably developed from the agitated running around with scent marking of primitive bees, to the pointing, if you will, of dwarf bees, and finally to the transposed pointing of the honey bee in the dark of the hive, in which gravity is used to represent the sun's direction. Much more work must be done, however, before we fully understand the evolution and capabilities of this elegant communication system.

To escape from enemies and to find food are certainly essential activities of animals. Perhaps, it would seem, an individual could do these best if it were by itself. Yet this is rarely the case. Instead, from the simplest to the most complex, animals live in some relationship to their fellows. These associations allow them to share foods, and to outrun or overwhelm their enemies. To accomplish these purposes, animals signal to each other in whatever channels are open to them. From simple silent flight of birds, indicating danger, to the elaborate dances of honey bees indicating food, the social life of animals is facilitated by communication.

6

Sexual Attraction and Recognition

ULTIMATELY the time comes for animals to reproduce, to make their contribution to the next generation. There are many animals that reproduce asexually; they merely replicate themselves, with no other individual being involved. But, for most animals, reproduction is sexual. In an almost infinitely vast world, considering the small size of most animals, a male and female must find each other, recognize each other as of the same species, and bring about the union of their reproductive cells. In these processes, communication reaches its highest development.

The first step in the reproductive process is the attraction of the sexes to each other. Communication signals are involved in this, and usually they are species-specific, so that when an attracted individual reaches the attracter, fruitful mating is possible. Attractive signals must usually act at a distance, preferably at as great a distance as possible; therefore, chemical and acoustical signals are most usually employed. Where animals are relatively close together, or in the darkness of night or ocean depths, visual signals may also be used.

Once the individuals are brought together, they must achieve final and absolute recognition of species and sex. The signals used for this may precede or may be linked to those involved in the next stage of reproduction, courtship. All senses may be involved in identification of species and sex, but vision is generally dominant in animals with good eyes, and the chemical senses in animals with poor eyes.

CHEMICAL SIGNALS

Probably the most primitive and widespread signals used for sexual attraction are chemical. These not only attract, but, since a wide variety of chemicals are available, also act as identifiers of species simultaneously.

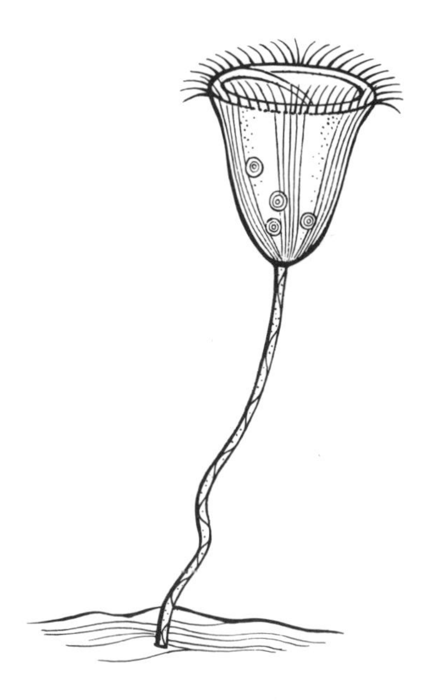

FIGURE 8. Vorticella, *a fresh water protozoan which attaches to algae or tiny bits of debris by means of the stalk. About* 250X *natural size.*

Protozoa

The Protozoa are the smallest of all animals (Figure 8), yet they intercommunicate. In the ciliates, there occurs periodically an exchange of nuclear material called conjugation. Two individuals come together and, by exchanging parts of their nuclei, get new genetic possibilities from each other. In some stalked ciliates, one conjugant remains attached to its stalk, while the other conjugant becomes motile. In this case the attached animal emits a chemical to attract the swimmer. This chemical can be effective up to 1 mm. away, over 250 times the size of the animal, equivalent, in man, to about 750 feet.

Spiders

In many species of spiders, the males use their chemical sense to find mates. Some species are wandering hunters, and, as the female hunts, she lays behind her a silken thread, the drag-line. This naturally carries her body odor. A male, on coming across one of these drag-lines, can detect whether it is from a female of his species or not, by using the extremely sensitive chemical sense-organs on his legs. If the drag-line is from a female of his species, the male spider next discovers the direction in which the female went by scent, taste, or touch, and follows the drag-line until he comes to her. Other spiders are web-spinners, and the females always stay in their lairs. The males must discover the species and sex by tasting or smelling the silk of the web, but, in this case, their search ends at that point.

Insects

Among insects, attraction of one sex by the other—usually the male by the female—by attractive odors, is almost universal. Even insects that use mainly other means for attraction and courtship —fireflies with their flashing, or grasshoppers with their songs— usually also use odors at some stage of the reproductive process. In many insects, odors are the only means for bringing the sexes together. We shall note just a few of the multitude of cases in which this happens.

The best known examples of attraction of males by females'

scents are found in the moths. Many species, possibly most species, are brought together this way. The larvae, or caterpillars, of moths include many economically important pests, and the hope that sex attractants might attract the adults to their death has led to a great deal of study on the attractants.

The female gypsy moth attracts the male by giving off an odor from scent glands near the reproductive organs. This chemical has been extracted from the glands and prepared in relatively pure form. When exposed in the woods, it is amazingly attractive to males. The same is true for the cecropia and promethea moths, adults of the American silkworms. Males of these large moths, which may measure 6 inches across the wings, can be attracted from distances up to 5 miles away by a female giving off her scent. Even papers on which this scent is absorbed are attractive to males at distances of a few miles.

The females of a little moth whose caterpillars live in many stored cereals, *Plodia*, also give off a scent to attract males. The scent is long-lasting, and the male's sensitivity to it is almost unbelievable. For instance, one investigator reports that a glass vial which held two female *Plodias* for only four minutes, even after being open for two days, induced courtship in males when it was held nearby.

Commercial silkworms, when adult, are moths which, through generations of rearing in captivity, have lost their powers of flight. The female gives off a scent which attracts the male and also induces him to start his courtship pattern. This chemical has been prepared in pure form; it is a complex organic material: hexadecadiene (10,12)-ol-1. A glass rod dipped in a solution having only 10^{-10} micrograms per milliliter of this chemical elicits whirring in 50% of the males. This is a fantastically low concentration (in English units equal to 1/80,000,000,000,000 oz. in a gallon of fluid), and would be totally imperceptible to man. Interestingly enough, when the ability of this substance to stimulate the organs of smell on the antennae was studied electro-physiologically, it was found that the male's antennae received it at these remarkably low concentrations, but the female's antennae were totally insensitive to it.

Butterflies, those day-flying relatives of moths, include some species in which the males have characteristic odor scales on the wings. These scales, called androconia, apparently give the males specific scents. The odors, however, do not seem to be used to attract the females from a distance. This is usually done by visual signals. After the males and females have come together, however, the scents may aid the female in identifying the male.

In a little parasitic wasp, *Habrobracon,* the male is attracted to the female by her odor. Filter paper on which a female has stood for a short time induces the male to court actively. This is true also of many species of cockroaches. A paper on which a female cockroach has stood, or on which extracts of a female's body have been placed, stimulates the males to court when held above them. It is an amazing sight to see a group of unmated male cockroaches stand perfectly still when a clean piece of filter paper is moved above them, but start vibrating their wings when another piece of filter paper on which a few females have stood is presented. The sex attractant of the female American cockroach has been isolated and identified; it is another complex organic compound: 2,2-dimethyl-3-isopropylidene-cyclopropyl propionate.

In bumblebees, the males attract the females with a scent produced by mandibular glands near the mouth. They mark rows of trees and bushes with this to create a scent-path. This is similar to the use of scent-trails to guide workers in many species of ants. In this case, however, the males guide adult females to a mating territory.

Vertebrates

Among the vertebrates, only mammals use scents generally for sexual attraction, although, as we have noted, many groups use scents for species recognition. Although it is known that some fish give off chemical substances which identify the species, or even the individual, there is little evidence that these are used to bring the sexes together. It is possible, however, when further studies are made, that cases of this may be found. For instance, females of some gobies, which are marine fish, release a chemical from their ovaries that attracts males and induces them to court.

In many species of mammals, the female gives off a scent when she is ready for mating, the rutting or oestrus season, and this attracts the male. Even where attraction is by some other means, recognition of the species in most mammals is effected mainly by detection of the odor of the female by the male. Some mammals use the sense of taste also for identification. The male giraffe, for instance, tastes the urine of the female, apparently to determine her sexual condition.

In some mammals, the males give off characteristic scents, and these attract the females or allow the females to identify the males. The male chamois, for instance, marks trees in his territory by rubbing them with his antlers, at the bases of which are scent glands. The antler rubbing is in preparation for combat of the males, but, at the same time, the scent glands are brought into action and produce a trail, something like that of male bumblebees. The females are attracted to these particular regions, and are thus brought to the males.

ACOUSTICAL SIGNALS

The two most highly evolved groups of animals, arthropods and vertebrates, have developed excellent mechanisms for receiving sounds. Correlated with this, they have also developed elaborate systems of sound signals which are used for attraction of the sexes and identification of the species.

Crustacea and Insects

Spiny lobsters produce scratching sounds by rubbing filelike organs at the bases of the antennae against the carapace. When recordings of these are played back to the animals, the males answer vigorously. In addition to acting as territorial markers, these may also be signals which are used by males to call in females, but proof of this is lacking.

Insects, always the most versatile of invertebrates, show the widest variety of sound-signaling systems for reproductive purposes. Insects produce sounds in many ways, some of them apparently incidental to other activities, but many of them distinctly purposeful. Among the methods which are used for sexual signaling, we may mention the following: bumping a part of the body against an object, to make tapping sounds; snapping taut membranes, the tymbals, to produce a rapid series of clicking pulses, which fuse into a sound; and using stridulatory organs, filelike structures over which teeth are drawn, to generate trains of pulses.

The tapping sounds that are produced by insects can be transmitted by air, which is the way we hear them, but the insects more usually receive them as transmitted by wood or soil. A solid material is a much better conductor of sound than air, and many insects have sensitive sound-receiving organs in their legs. Tymbals and stridulatory organs produce air-borne sounds of many types, and these are received by special hearing organs. The sound-patterns produced by insects and the ways in which sounds are used in mating are extremely varied. We shall mention only a few examples here.

It is often difficult to know, when studying insect behavior, whether sounds that insects produce are meaningful to the insects or not. Whenever an insect flies, the movements of its wings through the air are bound to produce sounds. If the wings move very rapidly, as they do in small flies and bees, then, to the human ear, the sounds resemble musical notes. Flight sounds, of course, could be merely incidental, an accompaniment to flight and nothing more; however, this is not always the case.

Female mosquitoes attract males by their wing sounds. The sounds of different mosquitoes are distinct, because the wings of different species are of different size and form, and the movements vary with the species. Males receive the sounds through highly specialized hearing organs, called Johnston's organs, at the bases of the antennae. On receiving the sound produced by a female of his own species, the male finds the female and mates with her. Male mosquitoes can be induced to fly to, and attempt to mate with, a small loud speaker or a tuning fork producing the proper sound. It is thus that tiny mosquitoes, ranging gigantic territories in comparison with their size, are able to get together and identify their partners.

Many species of beetles produce sounds by bumping some part of their hard bodies against an object on which they stand. The most famous of these are tiny, wood-boring beetles, named, ominously, death watch beetles. The adults strike some part of the body, probably the head, against the wood on which they stand to make a sound like the ticking of a watch. Interestingly enough, many of the early observers induced them to make the sound by placing watches near them. Apparently the beetles mistake the ticking of the watch for the sound of another member of the same species, and answer it. The sound is probably a sexually attractive signal, for these beetles have been observed to answer and approach another individual that is ticking.

Among the most persistent and interesting of all the sound-producing insects are the cicadas—often called locusts in the United States—and the crickets and grasshoppers. Cicadas are noisy, daytime musicians, the male alone singing. The sound is produced by snapping a special structure, the tymbal, with a muscle. In simple terms, it is like pulling in the bottom of a tin can to make it snap. As soon as the structure is snapped in, it is released, and it snaps out, only to be pulled in again. The tymbal muscle can contract 200–500 times per second, so that the sound pulses come far too fast for the human ear to distinguish them, and we hear a rattling sound. The female cicada, however, has quite a different ear from ours, and receives the snapping as pulses. The pulse-rate seems to be the specific feature of the song, allowing the female to distinguish the species and inducing her to fly to the correct male. While singing, the male is deaf, as a result of the action of a special muscle which folds the tympanum of his ear as soon as the tymbal muscle starts. Thus his delicate ears are protected against his raucous song.

Among the most studied of singing insects are the grasshoppers and crickets. There are two major groups of grasshoppers: the short-horned grasshoppers (Acrididae), which are most commonly seen jumping about in the grass in the daytime, and the long-horned grasshoppers (Tettigoniidae), often called katydids or false katydids, which are not too commonly seen, for they are usually nocturnal.

The short-horned grasshoppers—the word, horn, here refers to the antennae—make sounds generally by rubbing the hind legs, which are usually armed with comblike scrapers, against the wing covers, causing the latter to rustle. By varying the rhythm of

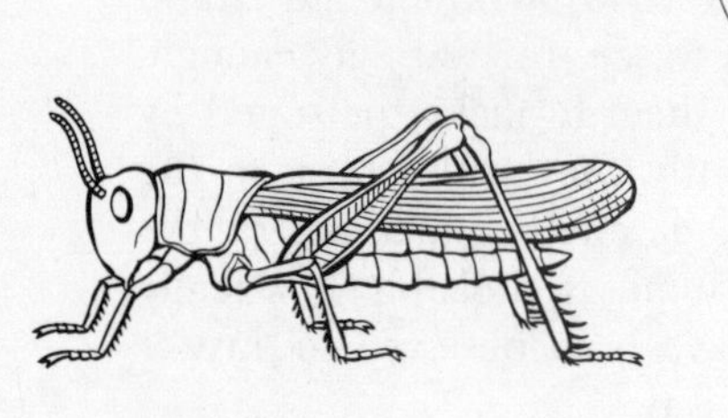

Short-horn grasshopper

Long-horn grasshopper

FIGURE 9. Top (left): *Short-horn grasshopper*; top (right): *long-horn grasshopper. About natural size.* Bottom: A *male katydid, showing the organ at the wing bases with which it produces sounds.*

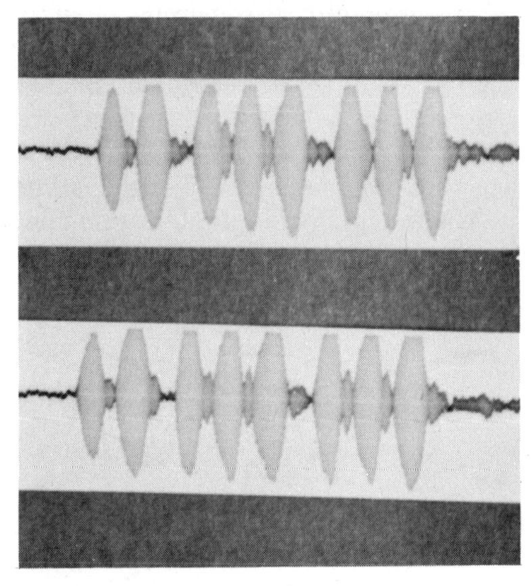

FIGURE 10. *Visible patterns of the song of a male tree cricket. Note the pulses in a regular sequence; these convey the information.*

movement, they produce a wide variety of sound-patterns. The long-horned grasshoppers make sounds by rubbing together a file and tooth or two files on their wing covers, causing the latter, or special areas in them, to vibrate. The songs of the short-horned grasshoppers are generally rustling, low intensity noises, and are usually heard in the daytime. The songs of the long-horned grasshoppers are complex sounds, sometimes sharp rasps, as in the katydid, sometimes high-pitched, almost pure tones, as in some meadow grasshoppers, and are usually heard at night. In both types of grasshoppers, the sound patterns of different species vary in the arrangement of the pulses in time—some chirp, some produce regular pulses of sound, some mix pulses with whirring noises, and some produce a series of bursts, then stop, and then produce another series of bursts. To the human ear, many of these songs are distinguishable by their frequencies. This, however, seems to be relatively unimportant to the insects, for their ears respond mostly to the pulsed time patterns of the sounds.

Crickets, like long-horned grasshoppers, produce sounds by rubbing together stridulating areas on the wing covers, using a rapid fluttering motion to produce a typical vibrato chirp. Some species sing, as do the long-horned grasshoppers, at night; some species sing, as do the short-horned grasshoppers, during the day; and others sing both day and night. By singing at different times of day, more than one species can use the same pulsed codes for their messages without confusion.

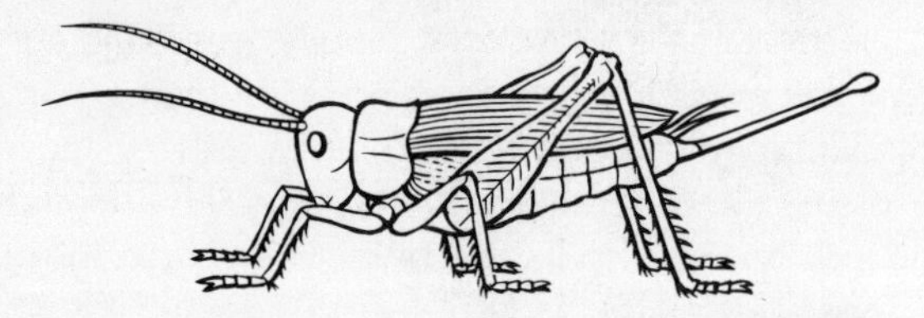

FIGURE 11. *Female cricket. About natural size. The female is the reactor, not the singer.*

The cricket or grasshopper song most usually heard is that of the male, the ordinary or calling song, to which the female is attracted. Although in most species only males sing, there are some types of short-horned grasshoppers in which the females can also produce sounds, and they answer the males. Each species has its own characteristic song pattern, and singing insects can be as easily identified from this, even by man, as from anything else. The females generally respond only to the call of the male of their own species, so species identification is made essentially before the sexes meet.

Insects that sing have highly complicated hearing organs, with an eardrum-like structure, the tympanum, to which the sensitive nervous elements are attached. Short-horned grasshoppers have their ears on the side of the body, where the abdomen joins the thorax. Long-horned grasshoppers and crickets have their ears on the front legs. There are some advantages to the latter, for the ears can be moved far apart to increase their directionality. These ears respond most when sounds come at them unevenly from the two sides. If the sounds come from directly in front, the ears respond only very slightly, and the insects may not be able to hear them at all. Thus the female, on hearing a male calling, can literally "zero in" on the male, for when her legs are facing directly toward him, she may no longer hear him, and she starts to march toward him.

Ordinarily male grasshoppers and crickets begin to sing when they become mature and can sing until they die. In some species of crickets, however, the males cease singing after mating and start again only when they produce a new sperm-packet to be transferred to the female. Generally, females respond only when they have not been fertilized. In many short-horned grasshoppers, the development of fertilized eggs produces a large mass of reproductive tissue, almost completely crushing the abdominal air-sacs, which form an essential element of the hearing-organ. The female is thus rendered deaf until she has discharged her eggs, and she does not respond to the singing of the males.

The fact that insects do not have a constant body temperature, as do man and higher vertebrates, creates a problem in their reproductive signaling. The movements of the wings in sound production are carried out by muscles, and muscles work more rapidly at higher temperatures than at lower ones. Therefore, the males sing faster at higher temperatures. For example, an American grasshopper, *Neoconocephalus ensiger*, almost doubles the rate of singing for every 10°C. rise in temperature. Obviously, if the female recognizes the species on the basis of the number of pulses per unit time, which she does, her nervous system must correct for temperature differences in the code. It has been shown for tree crickets that this is the case: females at 25°C. do not respond to the calls of males at 15°C. This is undoubtedly true, also, for other species. In nature, of course, this is no problem, for both sexes are at the same temperature—that of the environment.

Vertebrates

Among vertebrates, three large groups come to mind immediately when one thinks of acoustical communication in reproduction: frogs and toads, birds, and mammals. However, members of all groups of vertebrates can produce sounds, and these may be used in signaling prior to and during reproduction.

Some fish, such as drums and puffers, when removed from the water, produce bumping or clucking sounds. Only recently, now that it has become possible to record underwater sounds and later play them back to the fish, have we come to learn that these sounds are used as signals. Small, freshwater minnows, *Notropis*, produce sounds during mating. The male makes a characteristic thumping noise, and this is related to his defense of territory and attraction of the female. A number of marine reef fish produce sounds, and, since the males answer recordings of these sounds played back to them, it seems possible that they are used in territorial marking and in mating. The difficulties in making underwater studies of this type have slowed up the work in this field.

Generally, reptiles are not vocal and noisy animals, and they do not use sounds for communication purposes. However, some reptiles can produce special sounds. Male alligators and crocodiles, for instance, during the mating season, attract females by roaring. The wall lizards, or geckoes, of tropical regions, are known to all residents of the area for their ability to produce loud barks at night. Although exact studies on the use of these sounds is still lacking, it is probable that they are used as territorial or reproductive signals.

Throughout most of the world, during certain seasons of the year—in temperate climates, the spring and early summer—the nights are enlivened with choruses of calls coming from ponds where frogs and toads have come to breed. There are three major groups of tailless Amphibia: toads, frogs, and tree frogs. The males come first to the breeding pool, and set up the chorusing that attracts the females. Toads typically make rattling or whirring sounds; frogs emit chugging or croaking sounds; tree frogs produce whistles or continuous peeps, often sounding like insects. Members of different species are segregated, in part, by the fact that they have particular breeding patterns; thus, toads usually prefer temporary pools, whereas frogs usually prefer permanent ponds. The species are also separated by the ability of the females to identify the correct males from their songs.

As with insects, males produce specific calls, to which the females respond specifically, and thus the sexes come together. The calls are usually so distinct that man too can distinguish the sounds and use them to identify the species. As with the insects, furthermore, it is probable that some of the features of the signals which seem most important to man are of little importance to the females.

The calls can vary in a number of different ways: in sound frequency, sound quality, intensity, duration, and repetition rate. If different species use the same breeding spot, their calls usually differ in at least two of these qualities, making them quite distinct. Of all the characteristics of the song, as with the insects, it is the rhythmic pattern that seems to be the most specific.

In addition to calling in the females, the males' songs induce other males to answer. Thus, breeding choruses are set up, and the loud sound travels much farther than if only one male were singing; this attracts more females. The breeding chorus is usually not a random affair, with every male singing on his own. Male tree frogs of some species, *Hyla crucifer*, for instance, sing in trios, the three males singing in a set order. The total breeding chorus is a multiple set of these trios. The chorus may, at first listening, seem to have little or no pattern, but when one listens carefully enough to discover its intimate structure, it is surprisingly complex.

All these facts about the calling of frogs and toads—that the calls are specific, that only males call, that there is organization in the choruses—suggests that the calls are used in mating. This seems so reasonable that it has been accepted by almost all students of frog and toad behavior. However, only in recent years has proof of this been obtained.

It was not until 1958 and 1959 that female frogs and toads were actually observed by scientists to be attracted to males in the field, and, more important, to recordings of voices of the males, thus eliminating the possibility of visual or odorous attraction. With this proof of the place of the calls of frogs and toads in sexual identification—and thus in evolution of the Amphibia—research in this field is increasingly active.

Among the most familiar of animal sounds in the daytime are the songs of birds. These have attracted man's attention from time immemorial. Only recently, however, have we come to realize that bird songs are not expressions of joy, as they would seem, but are signals that male birds use to mark their territories and to attract females. The females of a few species of birds sing, but mostly, as with the amphibians and insects, the males do the singing, and the females are attracted.

Here too, as in the case of frogs, toads, and insects, the song patterns of different species are distinctive. Field ornithologists have long used the songs of birds to identify them. The development of the tape recorder has now made it possible to collect bird songs and to study them scientifically.

The songs of birds range from simple chirps or clicks to highly complex musical productions. The House Sparrow and Mourning Dove have simple songs, with little variability. On the other hand, many of the thrushes use a set of themes in kaleidoscopic recombinations to form dozens of different patterns. Some birds, such as Mockingbirds, incorporate into a basic song pattern of their own snippets and snappets from other bird songs, or even from human music. Carried to a greater extreme, some birds, such as mynahs and parrots, may desert their own vocabulary to imitate the vocabulary of man.

The ability of many birds to vary the combinations of simple themes enables them to develop individual features within the species. Each male American Robin of a group in one area may have his own particular style of singing. The females are thus able to identify not only the species, but also their own particular mates. Much of the individual character of bird songs apparently comes from the ability of birds to copy the peculiarities of others. Thorpe and his students in England have shown that male Chaffinches, reared in isolation, so that they cannot hear other members of their species, produce only an abstract, so to speak, of the usual song. As soon as they hear other males, however, they imitate them, and the song becomes fully developed.

Part of the song of birds, therefore, is fixed by genetic background, and part is learned as the individuals grow. However, species differ in this respect. Some birds inherit most of their song and are relatively unaffected by experience. Others learn most of their song from neighbors. Since the patterns of bird songs determine the pairing of the sexes, and can thus separate groups within the species, they are of primary importance in the evolution of birds. Newer techniques for recording and analyzing sounds should encourage studies in this important field.

Although most mammals, as we have seen, use odors for attraction and identification of the sexes, others use sound. We shall mention only a few cases. Anyone who has been rudely awakened by the serenading of a mate-conscious tomcat outside his window is well aware of the importance of sound for this mammal. Its larger relatives, lions and tigers, use sounds similarly. Many of the ungulates also use sounds to bring the sexes together. In moose, the males call to attract females; in domestic cattle the females bellow to the males.

OPTICAL SIGNALS

Optical signals are extremely important in identification and courtship, but are not usually used for attraction, except at very short distances, or in the dark. Generally, the use of visual signals for identification implies the development of good eyes, and therefore visual signals are common in higher animals, such as arthropods and vertebrates.

Marine invertebrates

Many marine animals, particularly those that live in the depths of the ocean, or come to the surface only at night, produce light. The light-organs, photophores, are often highly developed and, like eyes, have concentrating lenses to increase the brightness. Furthermore, the animals may flash in characteristic rhythms. It would certainly seem that these light patterns must have some purpose, such as sexual attraction or identification, but, in general, this is not known.

In a marine worm, *Odontosyllis* (Figure 12), males and females come to the surface at night during the breeding season and both sexes produce light, but with different patterns: the females glow continuously; the males flash. The females generally get to the surface first and form large, glowing masses, which attract the males. If the females fade before the males get there, the males stop and flash. When the females light up again, the males resume their course and mating occurs. This is one of the few cases, however, in which evidence is at hand that these lower animals use this well-developed talent for sexual communication.

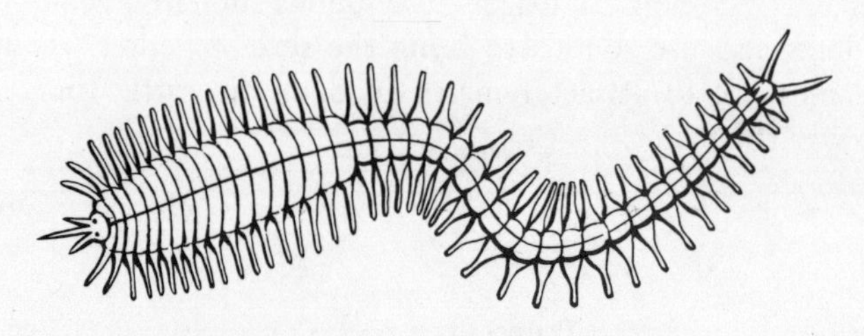

FIGURE 12. *A marine worm,* Odontosyllis, *showing two types of body segments. About ⅓ natural size.*

Insects

In some insects, the males display their bright wings to females to attract them. Obviously, even with the well-developed eyes of many insects, the signals can act only at relatively short distances. Male damsel-flies and dragon-flies pirouette, in aerial displays over the water, to attract the females. Male dragon-flies have plain wings, but male damsel-flies have brightly colored wings, which make flashing patterns in flight. These displays attract the females and act as species identification, for the females only respond to particular flight patterns.

Butterflies generally have bright patterns on their wings, and these are used by males of many species to attract females. Male butterflies may flash their distinctive wings while at rest, or fly up into the air to display them. Sometimes the flight of the males is also accompanied by sounds produced by snapping the wings against special parts of the body.

An interesting type of visual attraction, which is in a sense not really communication, and yet involves it, is the swarming of many species of small flies, such as mosquitoes and gnats. In this case, flying clouds of males and females perform an aerial dance above prominent features of the environment, such as trees or bare areas among grass, called swarm markers. In some species, the swarms consist of males only, and apparently they attract the females visually. In others, both males and females join the swarm. The males of most of these small flies have bushy antennae, with well-developed Johnston's organs, as in mosquitoes, and they probably identify the females by the sound of their wings. Thus, the original sexual attraction may be visual, with the females as selectors, but final identification is auditory, with the males discriminating.

Just as animals that live in the deep sea, where there is perpetual darkness, may use light to attract others, so terrestrial animals that are nocturnal may use flashes of light for attraction. The most famous of these are the lightning-bugs, or fire-flies. Actually these are not bugs or flies at all, but small beetles. In many species, the females cannot fly and are called glow-worms. These are not the only luminous insects, by far, for many species of beetles and the so-called lantern-flies also glow in the dark. A rather showy luminous beetle is the railroad-worm, which has a red light on one end and a green light on the other.

Although the color of light varies with different species, the identifying feature for most fire-flies is not color, but the rhythm of flashing. In general, both males and females flash in specific patterns. In one American species, *Photinus pyralis,* the male flashes more or less rhythmically, but the female flashes only upon seeing the flash of a male. She then flashes exactly two seconds after the male has finished his flash. If she does this, the male recognizes her as a female of the right species and comes to her. One can attract a male of this species experimentally by flashing a light two seconds after the male finishes his flash. In this case, the timing of the signal is most important.

The males of some tropical fire-flies join in light-choruses—if we might call them that—that is, the males flash synchronously, making imposing displays, with hundreds, or thousands, of males covering a tree and flashing in unison. This has been known for a long time, but how the fire-flies manage to synchronize is still unknown. It is also a matter of speculation as to what particular function this serves. Perhaps it acts, as the chorusing of frogs and toads, to build up the intensity of the stimulation, and thus to bring in females from greater distances.

How these insects produce light, essentially without heat, in measured time parcels, has interested not only the inquiring biologist, but also the chemist. It is known that the light is produced by the action of an enzyme, luciferase, on a substrate, luciferin, and that this requires oxygen and water. Many details of the process are known, and the chemical structure of at least luciferin is known, but not luciferase. We cannot manufacture cold light yet, and thus mimic these fine optical signalers.

Vertebrates

In turning briefly to the vertebrates, we find that visual signals are usually used for identification only; attraction involves chemical or auditory signals. Thus, in fish, many species have brightly colored patterns or specially colored spots, and these are used, when the individuals catch sight of each other, to identify the species.

Birds use their songs for both attraction and identification. However, they do not depend on these entirely, and final identification occurs during courtship. This usually involves discrimination of the brightly colored patterns of the males, for these are species specific. Some mammals, likewise, use combinations of body patterns and movements to make final identification of the species, but most use species odors.

All the distance senses, but especially the chemical and acoustical senses, are used by animals for sexual attraction and identification. The particular virtue of chemical signals lies in their persistence and the wide variety available. The special virtue of acoustical signals lies in their ability to travel for long distances. Visual signals can only be used for attraction at a distance if the animals live in the dark and can use flashing patterns. They may, however, be important for identification, in animals with well-developed eyes, when the sexes are close together. Attraction and preliminary identification by these senses is succeeded, in the reproductive ritual, by final checking of the species, and then courtship, in preparation for mating.

7

Courtship and Mating

WHEN members of the opposite sex have come together, whether accidentally, which is rare, or by using special signals, final recognition of sex and species is necessary. This, however, is not enough. The act of mating is a complicated process and usually requires extreme coordination. Rape is very rare in the Animal Kingdom. Male and female are usually brought into the proper state before mating. To arouse the mating instinct, and finally to achieve copulation, many animals perform elaborate courtship rituals.

In most cases, the senses used in courtship are those which are best developed in the particular animal, and the courtship movements are derived from nonreproductive behavior patterns. Most of the less specialized invertebrates have little need for courtship patterns; the use of specific attractants identifies the species and mating proceeds immediately. Higher animals, however, interpose many steps between sexual maturity and the final union of sperms and eggs.

ANNELIDS

A few annelids show what might be called courtship patterns. In common earthworms, at appropriate times of the year, individuals come out of their burrows and seek others. Each earthworm is both male and female, an hermaphrodite, but generally it does not fertilize its own eggs; it obtains sperms from another worm. Obviously no sexual identification is necessary, but species identification is. How this is carried out is not known. Earthworms have good chemical and tactile senses, and these could undoubtedly be used in the darkness, in which mating occurs. When two worms find each other, they join their bodies by the ventral surfaces, with heads facing in opposite directions. A special mucus-secreting gland, the clitellum, then secretes a slimy band that binds the two worms together, while they exchange sperms. Little is known about the senses involved in this, but communication between the two individuals is undoubtedly used to coordinate so complicated a process.

Most leeches, blood-sucking annelids, live in fresh water, but the land leeches lurk in water held in the bases of leaves of tropical plants. When ready to exchange sperms, the land leeches tap the leaves with the front end of the body to attract others. They then entwine variously before mating. It is obvious that this must be tactile or vibratory communication, allowing identification and correct positioning of the leeches, so that mating can occur.

It is unfortunate that we know so little about courtship and mating patterns among most of the invertebrates, and particularly the annelids. Many of these animals, however, are small and secretive, carrying out reproductive functions at night, and it is by no means easy to get information about them.

MOLLUSCS

Among the molluscs, the most highly developed are the cephalopods: squids and octopus. The male squid or octopus has specially developed tentacles with which he transfers the sperm packet, or spermatophore, to the female. Some males also use the suckers on these arms for aggressive or courtship displays. Most of the cephalopods can change their colors with amazing speed and versa-

FIGURE 13. Upper: *Octopus; vary in size from about this to many times this size.* Lower: *Cuttle-fish; about ⅓ natural size. The octopus crawls about among rocks along the sea-shores; the cuttle-fish swims in the ocean.*

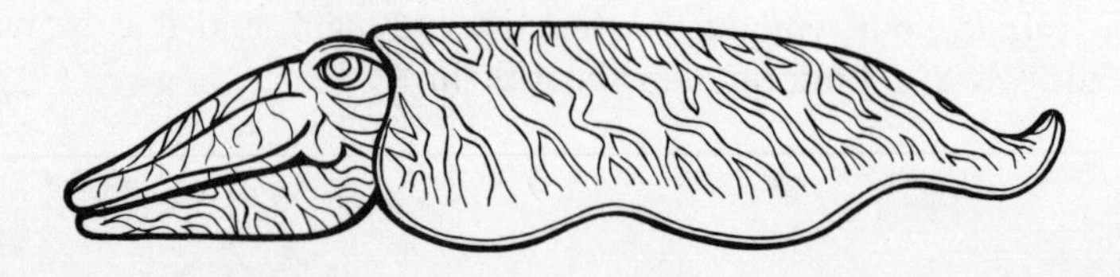

tility. It is a thrilling sight to see the colors flashing across the body of an octopus when it is excited. One would expect that the ability to make these rapid color changes might be associated with some communicative systems, but as yet we do not know much about it.

It is known, however, that, in some cephalopods, the color is used in courtship. In *Octopus horridus*, a small coral-reef-inhabiting species (Figure 13 upper), the male apparently identifies himself to and courts the female by sporting vertical stripes instead of his usual horizontal stripes. While the female is intrigued by the vertical stripes, the male implants a sperm packet inside her mantle with his tentacle.

In the cuttle-fish, *Sepia* (Figure 13 lower), the male adopts the so-called zebra-striped pattern when displaying before a female. This undoubtedly allows the female to recognize him and excites her sexually, for she allows him to insert a sperm packet with his special tentacles. The strongly optical trend in these mating patterns is correlated with the fact that the cephalopods have well-developed eyes, the best in the invertebrates.

The terrestrial slugs, on the other hand, have poorly developed eyes, but well-developed tactile and chemical senses, and they use these. Slugs seem to be superficially like earthworms, that is hermaphroditic, but the situation is not the same. In slugs and their close relatives, land snails, the animals are alternately male and female. However, males do not mate with females, but mate with other males, exchanging sperms with them. After the exchange of sperms, the individual that has been a male becomes an egg-producing female, and uses the sperms received from the other individual to fertilize them.

There seems to be no special attractive signals in slugs. When a slug is ready to exchange its sperms with another, it becomes

FIGURE 14. *Slug. About natural size.*

more active and so reaches, by accident, another "male" in the same condition. If they are both of the same species, the slugs start a courtship procedure, usually involving the mucus or slime trails which they produce at all times.

In the slug (Figure 14), *Deroceres reticulatus*, for instance, the individuals first crawl in circles, laying down mucus as they go. The circles become closer and closer, gradually becoming ovals, and finally the slugs are moving around with their right sides together. At this point, each of the slugs extrudes, from behind its right tentacle, a special organ called the sarcobellum, which is erected by filling with blood. This is used for mutual stroking and exciting. At the same time, they feed on each other's mucus, as they continue to crawl about. After a short time, they transfer sperms, using the sarcobellum as a conduit, and then separate.

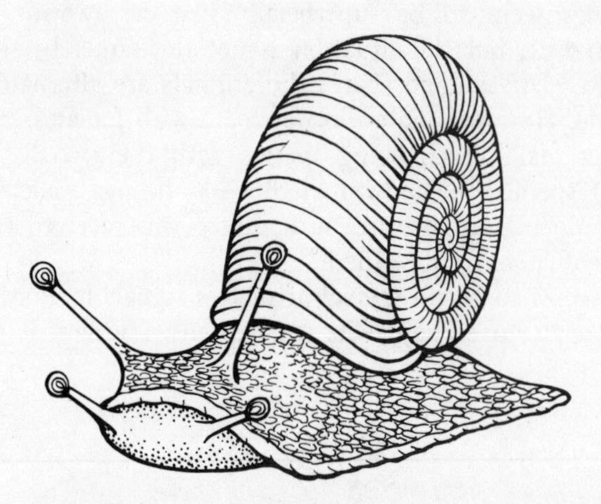

FIGURE 15. *Land snail. About twice natural size.*

In another species of slug, *Limax maximus*, there is no sarcobellum. The courtship is much simpler. The slugs merely crawl in a circle, the circle gradually becoming tighter, feeding on each other's mucus. From this mucus they form a thread, and on this they dangle in mid-air while mating. These are only two examples of mating patterns reported for slugs, which might seem to have little need or inclination for courtship.

Land snails (Figure 15) also exhibit a wide variety of behavior patterns for exchange of sperms between two "males." The most interesting of these is found in the European edible snail, *Helix pomatia*. In this case, individuals ready to exchange sperms find each other by random crawling, and, by tapping and joining their lips in a kiss of sorts, apparently identify themselves. Then one fires the so-called love-dart into the other. The love-dart is a small, sharply pointed projectile, which is produced in a special structure, the dart-sac. At first, the recipient of this rough love-gift recoils, but then the love-dart has an excitatory effect upon it, and it becomes receptive for the sperms of the other. Once the transfer has been completed, the second snail fires its love-dart, and the first snail becomes receptive. The love-darts apparently act as chemical communication mechanisms.

CRABS

Many crabs make sounds or have gaudy claws which they use in visual displays. The best studied of these are the fiddler crabs, so called because the males have one claw that is much larger than the other. Females, in general, have smaller claws, the two claws about the same size. Some male fiddler crabs attract the females by waving their claws at them, although often the meeting of the sexes seems to be more or less by accident. As soon as the males and females are within sight of each other, the males start courtship displays. The claw may be waved up and down vertically, or swung in and out sidewise. The rhythm and pattern of the display varies with the species. Even a human being can determine the species by the particular displays.

The males of some species of fiddler crabs make sounds by rapping the large claw against the side of their burrow, or on the surface of the ground. These sounds are also used to challenge other males, if they invade the territory. Although the basic pattern of courtship in fiddler crabs involves claw-waving, most species add other attractions. Some shuffle back and forth, some raise the body stiffly from the ground, others raise and lower the body rhythmically. The more complicated capers approximate the dancelike steps of other arthropods.

SCORPIONS

Scorpions and their tiny relatives, pseudoscorpions, have courtship rituals which have intrigued naturalists for a long time. The early observers compared the rituals to a promenade of the sexes, or, more fancifully, to a *pas de deux* in a ballet. Each species has its own characteristic rhythm and pattern of walking, so that species identification is effected. At the same time, the dance is necessary for the fertilization of the female. Male and female scorpions seem to find each other by chance, and possibly to recognize each other by odor. When they come together, both make aggressive displays, raising their abdomens as if to sting, and joining their claws. If they are of the same species and opposite sex, they hold together with their claws and shuffle rhythmically forward and backward. The male uses this movement to scrape and smooth a spot on the ground. On this he cements a hooked spermatophore. Then he pulls the female over the spermatophore; it hooks into her genital opening, and discharges the sperms into her.

Pseudoscorpions, which are much smaller than scorpions, have similar claws, but no poisonous sting. They behave similarly in courtship and mating. The male adds a special touch during the promenade by using his ramshorn organs, which are usually kept inside the body, but can be suddenly extruded to give him a horned appearance. It seems unlikely that these act as visual stimuli, for these animals have poor eyes. They might, however, give off a scent to identify the male and excite the female.

SPIDERS

Among arthropods other than insects, the most elaborate courtship patterns are found in spiders, whose sexual life is of almost unbelievable complexity and variety. Spiders have always fascinated man, and there have been many studies on their courtship and mating. We shall present only a few examples from this rich store of knowledge.

Problems of mating

Spiders have a severe problem in getting together at all, because both male and female are so fierce. They spend their lives hunting or trapping other animals, usually insects, which they kill with their poison fangs. There are two major groups: the wandering or hunting spiders, which spin no permanent web, but hunt their prey as do wolves and lions; and the web-spinning spiders, which build a web with sticky strands, into which insects blunder and are trapped. Hunting spiders generally have good eyesight, although some hunt only at night and depend mostly upon touch and smell. Web-spinners generally have poor eyes, but they compensate for this with an exquisite vibration sense, sensitive to the slightest twitching of the web; this enables them unerringly to find their helpless victims.

In most species of spiders, the female is larger than the male, for she is full of eggs and has large silk glands to produce a cocoon for the protection of the eggs. Her instincts are to seize any moving animal small enough for her taste and to kill it with venom from her fangs. So, were the male merely to walk up to the female, she would undoubtedly pounce upon him before he had a chance to carry out his reproductive function. It is essential, therefore, that most male spiders have some way to identify themselves to the females in time to escape sudden death.

In addition to having communication signals to identify themselves to the females, male spiders also have ways of protecting themselves when near such dangerous mates. The male spider is one of the few animals in which the organ used to transfer the sperms to the female is not near the testes, where the sperms are produced. In most animals, the sperms move through a tube from the testes to the outside of the body, and, if special organs are

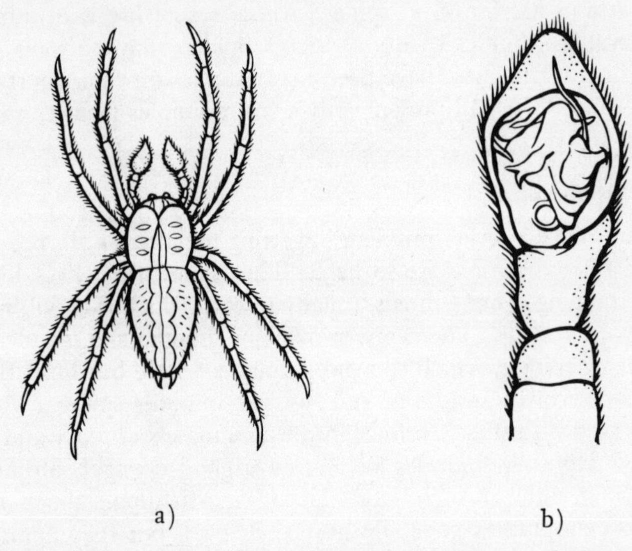

FIGURE 16. *a) A male spider of the wandering type, not having a fixed web, showing the enlarged palpi used for holding sperms. b) Enlarged view of end of palpus of male spider, showing the reproductive apparatus; about 15X natural size.*

used to inseminate the female, the organs are present at that point. In the case of the male spider, however, this is not so. He has the last segment of each palpus, an appendage of his mouthparts, developed into a special structure for transferring the sperms to the female. In its simplest form, this is like a medicine dropper, with a bulb into which the sperms are sucked through a tube, which can later be inserted into the female (Figures 16a and 16b). In its most complicated forms, designed to fit only into the genital openings of females of the right species, the organ is unbelievably complex.

When a male spider is mature, he spins a tiny sheet of silk on which he deposits the sperms from his abdomen. He then turns around and sucks the sperms up into his palpi. Here they are carried, ready for him to reach past the hungry jaws of a female, insert the palpal organ into the female's genital opening, and fertilize her. In most cases, the opening of the female reproductive organ is like a lock, and the tube of the male palpus like a key. Even if a mistake were made during the courtship, a male spider could not introduce his sperms into the wrong female. Mating in spiders is so hazardous that safeguards are present up to the moment of copulation to assure that sperms are not wasted on females of the wrong species.

So, with the fluid which will transfer the spider's life to the next generation in special appendages where it can be carefully guarded, and from which it can be transferred to the female with a minimum of danger, the male sets out to find a mate.

Force and stealth

In a few species, the male spiders are as large as, or larger than, the females. In this case, there is little problem. It does not follow, however, merely because both sexes are of about the same size, that there will be no courtship. It is not necessary, however, and some do not show it. In some crab spiders, so called because they can walk sidewise like a crab, in which the male is larger than the female, mating is almost rape. The male comes upon a female, seizes her, and after some preliminary rubbing of her body, inserts the palpi and fertilizes her. In other crab spiders, in which the female is as large as the male, and thus a potential danger, the male, after rubbing and seemingly hypnotizing her, ties her with silk to the leaf upon which she is sitting. When she is tied down securely, the male proceeds to insert the palpi and transfer the sperms. She is then left to free herself.

Most spiders in which the male is about the same size as, or smaller than the female, cannot get away with such blunt tactics. However, some of them have stealthy solutions for the problem.

Male spiders of the genera *Drassodes, Zelotes,* and *Drassus,* which make nests under rocks and range out from them to hunt, capture females that are small and immature, and imprison them in silken prisons beside their own nests. Immediately after the females molt to adulthood, and are still soft and helpless, the males mate with them and then leave. This has all the elements of an old-time melodrama, but, biologically, it is perfectly acceptable.

Some male spiders try to appease the female's predaceous instincts by bringing her something to eat. The male *Pisaura listeri,* a little wandering lynx spider, catches a fly and wraps it in a silken ball. When he finds a female, he tempts her to dash at him, and when she does he hands her this prepackaged food. How could a lady resist such culinary convenience? While the female is busy feeding upon the insect, the male quietly transfers the sperm to her.

Vibrations and sounds

Most male spiders, however, do not get away so easily. In the case of web-spinners, the male is presented with the problem of getting into the female's web without having the vibrations he produces mistaken for those of a prospective victim. Web-spinning spiders are usually short-sighted, but they have a remarkable vibration sense. Therefore, the males of many species use the senses of touch, vibration, and occasionally hearing, to identify themselves.

Male web-spinners usually leave their own webs to search for females when they are mature and have filled their palpi with sperms. The discovery seems to occur mostly as a result of random search, but the identification definitely does not. Male spiders collect only at the webs of females of their own species. The identification depends upon their ability to recognize the webs, probably by scent or taste.

The web-spinners construct a variety of snares. The most familiar are the cobwebs, tangled masses of silk, frequently found in houses. Also quite familiar are the orb-webs, beautiful geometric patterns of silk that glisten in the morning dew, built by garden spiders. However, these are only two of a number of classes: funnel-webs, flat sheets with a funnel at one end; combined tangles and orbs; webs that have been called bowl-and-doily, because they look like a silken bowl sitting upon a silken doily; and many others. Since different species make different webs, the males of each species must approach the webs properly. We shall mention only a few examples.

Relatives of the hairy tarantulas, *Atypus*, find or dig holes in the ground, and there they make tube-webs, with part of the tube lying on the surface of the ground. If an insect ventures upon the tube, the spider lunges from its station inside, kills the insect with its fangs, and then drags it inside to eat it. The male spider, approaching such a vicious creature, must be sure to identify himself correctly before the fangs reach out and kill him. The male recognizes the correct tube apparently by its scent and taps it carefully with his front legs in just the right rhythm. The female apparently recognizes this rhythm, for she does not immediately seize the drummer with her fangs. If she has already mated, she shakes the web vigorously and makes aggressive passes at the male. The male understands the message and leaves. If she has not mated and is receptive, she plucks gently at the web to signal to the male to enter and live with her as her mate.

In many species of orb-weavers, the male identifies himself to the female by the rhythm of plucking of the web (Figure 17). He stays at the edge of the web and, with his front legs, snaps the strands, as a harp player plucks the strings. In many cases, the female responds to this as if he were a prospective item of diet, and rushes at him. The threatened male retires cautiously from the web, for the female will not leave it, and waits until she goes back into her lair. Usually such a reception does not deter the male, and he plays his vibratory serenade again. Ultimately, if the female is ready to mate, she remains quietly in the center of the web, and the male, plucking all the while, moves in—often on the other side of the web—and fertilizes her.

In some web-spinners, the male comes to the web and waits quietly until a fly lands on it; then, after the female has rushed out, seized the fly, and is busy feeding upon it, the male starts his plucking and slinking. It is as if he were waiting for her hunger

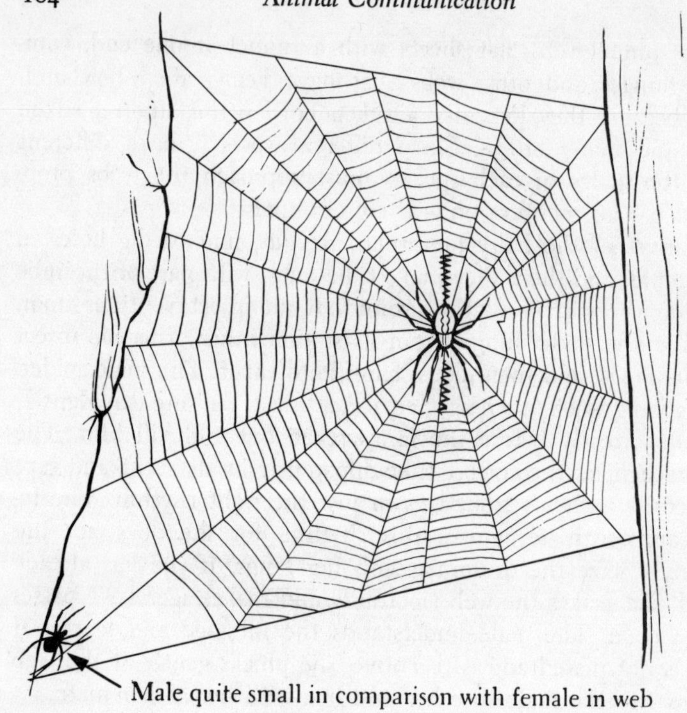

FIGURE 17. *Orb-web with female spider in center and male approaching at lower left.*

to be appeased before he starts on his business. There are many other vibratory displays used by male web-spinning spiders, but the above will serve as examples.

The impression should not be left that all female spiders are such ogresses. In some species, the female allows the male to remain nearby, or even to live in the web with her. Some web-spinning spiders construct their lairs close together, and the webs of males and females become intermeshed at the bases without any particular strife. The female house spider, commonest of spiders in homes in temperate regions, not only allows the male to live with her amicably in the web, she actually courts him. She plucks the web, causing it to vibrate, and he answers. When both are thus aroused to the reproductive state, she comes to him and excites him to mate with her. They may spend many hours, with the male inserting first one palpus and then the other into the female. Often she seems almost insatiable, and pursues the male long after he apparently has no more sperms to give.

Some species of web-spinning spiders use sound production for sexual communication. The male *Steatoda* produces sounds by rubbing a corrugated area of his abdomen against his cephalothorax. He also taps on the web to attract the female's attention. The female allows the male to live peacefully in her web in a special lair which he builds. When both are ready to mate, the male, through his tapping and stridulating, excites the female. Then he builds the mating bridge, a heavy line of silk, and, while face to face with the female on this, he introduces the sperms into her. After each introduction, the male stridulates vigorously, serenading his mate.

Some nocturnal hunting spiders build silken nests to occupy during the day. If a wandering, mature male finds the nest of a female of his species while she is not in, he builds a nest beside hers—a real "love nest." He is smaller than the female, and so his nest has too small an opening for her to enter. When the female returns to her nest, the male starts to drum rhythmically on the wall between. This procedure, which would arouse considerable distress between human neighbors, has the opposite effect here. If the female is unmated, she allows the male, who has thus identified himself, to enter her nest and mate with her.

Visual displays

The most elaborate courtship displays in spiders are found in the long-sighted, diurnal hunters. There are two major groups of these: the wolf spiders, whose name indicates their ferocity, and the jumping spiders, whose name indicates their special talent.

When a male wolf spider is ready to mate, and has charged his palpi, he sets out in search of a female. Since these spiders are wide-ranging, this could be quite a task. However, all spiders lay down behind them a silken drag-line, so when the male comes across the drag-line of a female of his species he need merely follow it to find her. If the odor shows that the silk is from the right species, the male races along following her.

Ordinarily the female wolf spider pounces on anything that she sees, and the male would be no exception, if he did not identify himself. Therefore, when the male overtakes the female, he attracts her attention, but stays far enough away so that she cannot reach him. Then he performs a special dance to identify himself.

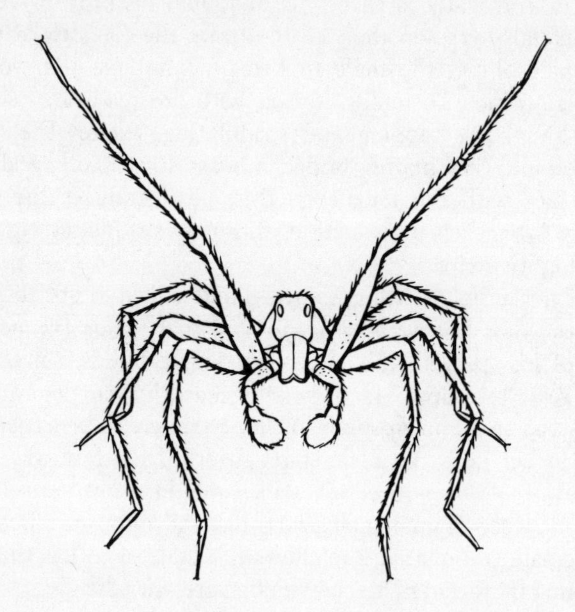

FIGURE 18. *Male wolf spider displaying front legs and palpi. About twice natural size.*

This dance is derived from his defensive attitude, in which the long front legs are thrown up in front. In many wolf spiders, the front legs are highly decorated with spines or hairs and, as the male's excitement increases, the spines on his legs are erected like the bristles of a bottle-brush (Figure 18). Furthermore, he moves his legs up and down in a regular rhythm, sometimes one leg at a time, sometimes both; every species has its own pattern. Some males also tap their legs on the forest floor, and one American wolf spider, *Lycosa gulosa*, is called the purring spider, because of its rapid tapping.

Courtship is often a protracted process, for the female is coy and persists in suddenly dashing at the male. He runs away nimbly, and starts over again. However, if she is in a receptive mood, she gradually seems to recognize the individual in front of her and lowers her defensively raised front legs. The male, keeping up his dance rhythm all the time, moves closer, until he touches her. Now the rhythm comes through to her tactilely, and she allows him to reach around her sides and insert his palpi into her genital opening.

Jumping spiders have highly developed eyes, which they use for gauging the distances of their agile jumps. Many male jumping spiders, although no more than a tenth of an inch long, are brilliantly marked, whereas the females are drab. As would be expected, the courtship displays of the males are usually visual *tours de force*. To be sure, a few jumping spiders do not court at all; the males and females apparently identify each other either by visual or odor signals and mate without ceremony. Most of them, however, have some sort of courtship.

The males seem to recognize the females not only by sight, but also by other cues. Thus, males court only unmated females, even though the mated state could probably not be told from mere appearance. Some also court before they can see the female, presumably using a well-developed chemical sense to detect her presence. This raises an interesting question. Why have male spiders not developed specific odors to identify themselves to the females? The chemical senses of spiders are well enough developed for this. No one knows the answer, but possibly spiders lack characteristic scents, because these would give them away to their prospective prey.

Each species of jumping spider has its own particular courtship movements. These involve raising and lowering the front legs, either both together or alternately, movements of the often brightly colored palpi and jaws, and crablike sidling movements around the female. In a few species, the males use their eyes to signal also, moving the pigment in the eyes back and forth to make them flash and sparkle. Altogether these tiny dancers put on an impressive show.

INSECTS

Almost all insects show courtship behavior and communication between the sexes. Again we shall cite only a few examples.

In cockroaches, the female's odor induces the male of the species to rustle his wings. Males of some species raise their wings and offer to the female the secretion of a gland beneath the wings. This is quite attractive to the female, and offers her a further means of species identification—by taste. While she feeds on the fluid, the male joins his reproductive organs with hers, and concludes the mating.

Male butterflies flash their brightly colored wings for identification and courtship. In some species, if the female is receptive, she raises the abdomen so that he can join his reproductive organs with hers. In other species, raising the abdomen signals that the female is not receptive, or has already mated, and the male stops courting. The nature of the signals involved in courtship displays by butterflies has been analyzed for some species. In general, the color patterns and movements of the wings stimulate the females' movement-sensitive eyes.

Among the most studied of insect reproductive communication signals are those of crickets and grasshoppers. The male cricket or grasshopper, by its ordinary or calling song, calls the female to him for mating. When she arrives, the male changes from this song to the courtship song. As with the ordinary song, each species has its own courtship song, so that there is a double safeguard against mating with the wrong species.

After the male has identified himself to the female by the courtship song, he may be allowed to copulate. In this case, many male grasshoppers produce a mounting song as they mount the female, and a mating song as they copulate. Furthermore, many of them make special sounds at the conclusion of the copulatory act. In some species, the females also join in producing sounds before, during, and after mating. It is a complicated system of communication that finally brings the lives of these little animals to fruition.

The German scientist, Faber, as a result of years of careful study of grasshoppers in central Europe, has classified the calls that he can distinguish with his ear. For the approximately sixty species he has studied, he has distinguished over 400 acoustical signals, most of them used in courtship and mating. In one species alone, he has identified fourteen different types of songs produced by the males.

Senses other than the auditory can also aid in reproduction, even in insects which have well-developed sound systems. Thus, male grasshoppers of some species, on seeing an approaching female, do not sing a courtship song, but vibrate in characteristic rhythms branches of the plants on which they sit. They are joined in this by the females. Many other species use the sense of smell for final identification before mating. The long-horned grasshoppers have long antennae, as their name indicates, and it certainly seems as if these would be involved in identification by smell. Male tree crickets, like roaches, have glands on their backs which produce a fluid for the females to feed on, and, while the females are busy at this, the males mate with them. The so-called crackling grasshoppers combine visual and auditory signals. The males fly up from the ground and flick their bright red or yellow wings, while they make a snapping noise to court the female on the ground.

VERTEBRATES

In turning to the vertebrates, we find a bewildering variety of courtship patterns. Again, we shall discuss only a few selected examples.

Fish

Fish may use their chemical senses for identification of species or sex, but these are seldom trusted entirely. Visual displays, in many species, take over for final identification and arousal. For instance, the male stickleback displays his bright red belly to the female, while almost dancing in the water. Male platyfish show off their bright colors while moving broadside away from the female. Male swordtails display their decorated tails by thrusting them toward the female. If one crosses the last two species, the male hybrid offspring inherit the parts of the courtship behavior piecemeal, and so do not show a consistent pattern that would identify them as one species or the other. They are, therefore, unable to mate with females of either species.

The males of many fresh-water fish have brightly colored spots on the gills or fins which they display in courtship before mating. A special use for these spots has been suggested in the case of the mouth-breeders. In these fish, the female, after depositing her eggs, takes them into her mouth, where they are fertilized. There she keeps the developing eggs and, subsequently, the live young. The male of some species has spots on the anal fins, near which the reproductive system opens to the outside; these spots are often called eye-spots, but they are not sensory. After the female has placed her eggs on nearby leaves she picks at anything that is spotlike, and so she gathers up the eggs. The male, in the meantime, makes himself conspicuous nearby. After the female has the eggs in her mouth, she still continues to pick at spots, but now the most available spots are those on the male's anal fin. While she tries to take them into her mouth, the male discharges the sperms, which enter the female's mouth and reach the eggs.

Visual displays are the most common in courtship of fish, but we are now coming to appreciate that many fish make sounds as well. The male cod, for instance, uses a combined visual and auditory display. He establishes a territory on the floor of the ocean by chasing away other males and advertises it by grunting and flashing. When a female enters the territory, the male goes into the so-called flaunting display, spreading his fins and flipping around, while he makes a grunting sound. The female quietly observes this, and, while she is still, the male prods her to the surface of the water, and there spawning occurs. Many other fish also produce sounds most actively during mating season, so they are probably important in reproduction. However, much further research is needed in this area.

Amphibia

Male salamanders of many species are brightly colored, and apparently display their patterns to court the females. Other species use special scents produced from glands on the head. In one species, for instance, the male faces the female, bends his tail forward alongside the body, and waves it vigorously to create a water current past his face, which carries his scent to the waiting female. This excites the female to follow him. He then deposits a sperm-packet on some object, and, by backing up, leads the female over it. She picks it up in her cloaca, thus completing the mating act herself.

Frogs and toads generally aggregate specifically, because the females react specifically to the distinct calling songs of the males. The singing males usually seize any moving object of appropriate size. If a male seizes another male, the captive emits a special release note, which causes its captor to release it. A gravid female remains quietly in the grasp of the male. Thus the male frog or toad is informed that he has a mate if the seized object is of the appropriate size and texture, meaning puffed out with eggs, and if it remains quiet. If the object is not turgid, or gives the release call, the male turns it loose, and starts calling again.

Reptiles

In many species of lizards the males are brightly colored, and display the colors to the females. The male western race runner, or canyon lizard, for example, raises his body off the ground and spreads himself vertically to obtain the maximum color effect. While the female watches this, he approaches and finally seizes her in his jaws. Gradually he maneuvers her into position with his jaws so that he can wrap around her and join his reproductive organs with hers.

Birds

Among the best known of vertebrate courtship displays are those of birds. Many books have been written about them. Generally, the displays are visual, sometimes tactile, and often accompanied by sounds.

In some birds, the males are much more brightly colored than the females, as in the case of the American Cardinal or Mallard Duck. These brilliant patterns are not only useful for visual identification, but also are used in courtship displays. The males and females of other birds look almost alike, as in the case of the penguins or gulls, so that colors alone cannot be used for identification. In this case, the birds use their voices, which may be different in the two sexes, or have elaborate courtship rituals, often called dances, with the partners behaving differently.

Possibly the simplest courtship pattern in birds is feeding or mock feeding of the female by the male. In this, as the female comes into the male's territory and is approached aggressively, she signals to the male that she will not fight by assuming a submissive posture, followed by solicitation of feeding, as if she were a baby bird. This, in some species, induces the male to feed her, in others to do mock feeding, that is, go through the motions without actual food. Usually this is accompanied by appropriate sounds, allowing final identification of the species, and, through continued attention of one to the other, arouses the birds to full mating pitch.

In other cases, male birds indulge in special antics to identify themselves to the females and to arouse the females' mating instincts. Thus, male Birds-of-paradise display their brilliantly colored tail feathers. Male Bower Birds clear areas of the ground, where they build bowers of vegetation in which to display and call. When the females are attracted, the males lead them into the bower to mate.

In penguins and albatrosses, the male and female closely resemble each other; for sexual identification, they indulge in elaborate rituals or dances. These involve bill fencing or snapping, characteristic sounds, and ballet-like steps, during which the birds bow and raise their heads in the air. Close observation usually shows that the male in these displays behaves differently from the female, even though at first sight one may not be able to distinguish this.

As in the case of territorial or attractive displays, birds exhibit courtship behavior in an almost infinite grade of intensity from minor interest to extreme excitement. It is here that the value of motions and visual signals for transmitting a continuum of quantitative information can best be seen.

Mammals

Observations on courtship in mammals are also numerous. As we have noted, attraction and specific identification in mammals are usually based on odor or vision, sometimes with auditory accompaniment. Courtship is usually strongly tactile and visual. We shall cite only a few examples.

Male elk and moose use auditory signals to attract females from a distance, but identify and court by visual signals—gaits and postures. On the other hand, pigs use mostly tactile stimuli—pushing and nuzzling—in their courtship. It should be no surprise that the male giraffe arouses the female by a procedure called necking, in which the giraffes entwine and rub together their long necks.

Some mammals show an almost aggressive spirit in courtship displays. Male cats and horses, for instance, may actually bite the females, but the females seem to be aroused by this abuse. In baboons, monkeys, and apes, lip smacking is a feature of courtship and copulatory behavior. An African scientist reports that he is able to tranquilize and interest many primates in zoos, or even in the wild, by imitating the characteristic lip smacking with which they accompany mating. When one considers the lip contact and smacking that these primates use in their mating procedures, it is probably natural that one of the commonest forms of sexual arousal in that other primate, man, is kissing.

To assure that the sperms of a male reach the eggs of a female of the same species, identities are checked and double-checked, before the final mating act. Communication signals used for attraction are usually species-specific. When the sexes have been brought together, there is generally a period of courtship behavior, in which final identification is made, and the individuals are brought to mating pitch and position. In many cases, the reproductive organs are built like locks and keys, so that even if a mistake is made up to this point, the final mistake cannot be made. All this assures that in the process of reproduction, probably the most important in the life of any animal, the proper eggs and sperms are brought together.

8

Parental Care for Developing Eggs and Young

THE final result of sexual attraction, courtship, and mating, is the production of fertilized eggs. These must develop through the embryonic stages, and the young produced from them must grow and develop until they ultimately become adults. Animals have as many different ways of dealing with their eggs and young as there are types of mating. It is well, therefore, to discuss these, before we ask how communication is involved.

RELATIONSHIPS OF DEVELOPING EGGS AND YOUNG TO PARENTS

The most familiar animals, birds and mammals, take some care of their young, but this is by no means the usual situation in the Animal Kingdom. Most animals pay no attention at all to the developing eggs and young. The majority of marine and fresh-water invertebrates—worms and snails, for instance—produce fertilized eggs and desert them. They may cover the eggs with jelly-like materials before they leave, but that is all. In many marine invertebrates, the sperms and eggs are merely released in the sea-water.

Most marine fish leave their eggs in the surface waters of the ocean, where they toss back and forth during development, and where the young fish fend for themselves. Some fresh-water fish leave the eggs to themselves, but usually they fasten them onto objects, such as leaves or roots. Salamanders, frogs, and toads lay their eggs in fresh-water, enclosed in jelly-like cases, and leave them and the tadpoles hatching from them to develop entirely on their own.

On land, earthworms, snails, slugs, and most terrestrial arthropods, including the majority of insects, leave their eggs and offspring to their own resources. These eggs are usually deposited in suitable spots, and are often protected by a cocoon or mucous egg case, but otherwise the parents give them no regard. Among terrestrial vertebrates, many reptiles—snakes and turtles, for instance—deposit their eggs in soil or sand, and desert them.

In all cases in which eggs and young are left on their own, little, if any, interchange occurs between them and their parents, and therefore there is no development of communication systems.

The second general method for dealing with developing eggs is for the parent, usually the female, to brood the eggs, either on or off the body, with the young independent after they hatch. There are a number of marine invertebrates that do this. Female crabs have a broad abdomen, whereas the males have a narrow one. The broad abdomen is used by the female to hold the eggs while they develop. Some relatives of the starfish hold the eggs either on the arms, or in the stomach, while they develop. The female octopus lays her eggs in a sequestered spot in rocks or coral, and there she guards and fans them until they hatch.

Some marine fish also brood the eggs off the body, depositing them in crevices among corals or other materials. There they defend them and ventilate them with their fins during development. The female sea-horse lays her eggs in a pouch on the male's abdomen, and he is left to carry them during incubation. Most fresh-water fish lay their eggs in a nest or on vegetation, and the male or female, or both, guard the eggs and fan them with their fins to supply oxygen.

A number of fresh-water invertebrates also brood their eggs. Some leeches carry the eggs around on the body, and many of these carry the young also for a short time. Crayfish glue their eggs onto appendages of the abdomen, the swimmerets, and swing them back and forth in the water to ventilate them. Some also carry the young for a short time.

In the terrestrial sphere, a few species of frogs and toads hold the developing eggs on the body and some have specially developed tissue on the back into which the eggs sink. Many terrestrial salamanders lay their eggs under rocks or stones and curl around them to guard them. Some blow up a foamy mass in which they place the eggs, then hold the eggs and foam around their heads. Some lizards and snakes also keep the eggs with them in a lair or nest while they develop. Finally, there are a few insects such as earwigs, that make a nest and take care of the eggs in it, until the young hatch.

The third relationship entails the care of both eggs and young, until the young are developed to some stage. Any such situation obviously involves some sort of social relationship, and rather highly developed systems of communication. The two groups of animals in which this type of behavior occurs are the social insects and birds. In the case of social insects, the colony may consist of thousands of workers, and one reproductive individual. The workers take care of the eggs and feed the young, until they become adults and join the colony. Among birds, the eggs are incubated by the male or female, or both, and the young birds are fed and taken care of until they can fend for themselves.

The fourth method for caring for the developing eggs is to retain them inside the body of the female for development. In some of these cases, the young are on their own from the time of birth, while in others the young are cared for. A large number of animals bear living young. Such simple invertebrates as sponges and coelenterates keep the eggs inside the body of the female, until they have become larvae. Many other invertebrates also have this ability. There are quite a number of insects that give birth to living young, but ordinarily take no care of the young. Anyone who has kept fish in an aquarium is familiar with the live-bearing guppies and their relatives; the eggs are held in the body of the female, but the young are not cared for, indeed may even be eaten by the parents. Mammals have a reproductive system that allows the developing young to derive their food directly from the mother while inside her body. After birth, the young mammal has a period of dependency, on the female parent at least, and is fed with her milk.

Obviously, where the eggs develop independently, or only the eggs are brooded and the young not cared for, little or no communication, except between brooding parents, is needed. Where the eggs and young are tended, or the young are reared by parents after birth, communication systems become complicated, changing during the development of the young, as the adult patterns emerge.

SYNCHRONIZATION OF SPAWNING

Even in animals that desert the eggs, communication may play a role. Sessile animals, which are fixed at one spot and cannot move about, can neither attract each other nor truly mate. Yet some means is needed to assure the union of the proper egg and the proper sperm. Usually these animals exist in large groups, so that females discharge their eggs near sperm-producing males. However, this does not necessarily assure that the eggs and sperms are present at the same time. A solution could be achieved by producing the reproductive cells continuously, but there are few animals that can afford such a waste of material. A more practical solution could be found by reproducing only seasonally, males and females developing eggs and sperms simultaneously. In this case, synchronization of spawning is environmental, and has nothing to do with communication.

Obviously, the safety factor in this could be increased if, in addition to having seasonal synchronization, and to having males and females relatively close together, some means of communication were possible, so that the females "know" when the sperms are present, or the males "know" when the eggs are present. Many species of marine invertebrates have such a system. In fact, it is probable that most sessile, marine invertebrates that discharge eggs and sperms into the ocean have some communication between the sexes.

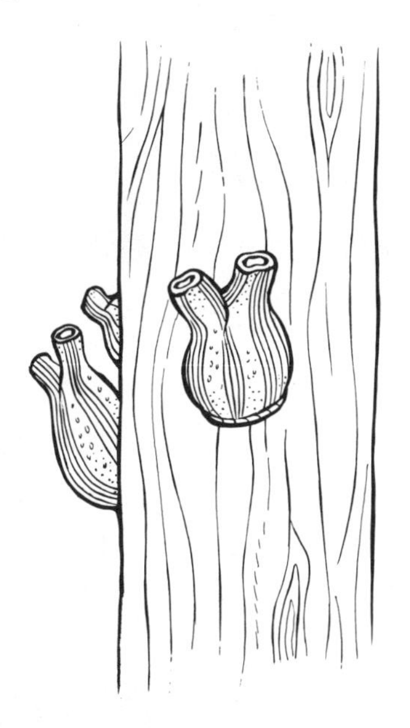

FIGURE 19. *Tunicates attached to a piling. These are common marine animals attached to submerged objects. About natural size.*

In some invertebrates, the presence of eggs in the water triggers off the release of sperms, or conversely the presence of sperms triggers off the release of eggs, depending upon which sex starts to release the sex cells first. In the mussel, *Mytilus,* the sessile tunicates (Figure 19), *Ciona* and *Phallusia,* and some sea urchins, either sex may release the reproductive cells first, whereupon the other sex responds by releasing its gametes.

In other species, the females release eggs spontaneously, but males release sperms only when stimulated by the presence of eggs. This situation is found in the marine annelid, *Platynereis dumerili*, and in a marine worm related to the tunicates, *Saccoglossus horsti* (Figure 20). Conversely, in the oyster, and in sipunculid worms, the male releases sperms first, and the presence of sperms in the water causes the females to release eggs.

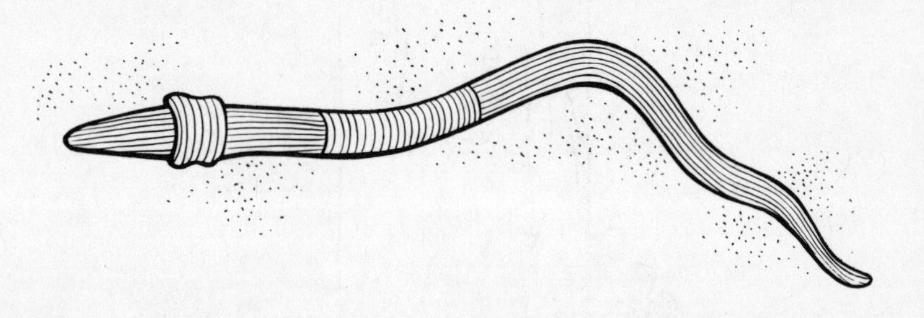

FIGURE 20. *A marine tongue-worm. About natural size.*

Thus, even in cases in which the parents do nothing about the eggs once they are discharged, communication is involved in a primitive way to assure that the correct sperms reach the correct eggs most efficiently. It is obvious that, since the eggs and sperms are released in the open sea, where eggs and sperms of many other species may be present, the actual selection, if we could call it that, of the correct sperms must be done by the egg itself. This ability resides in the membrane of the egg cell.

BROODING BEHAVIOR

When eggs are protected by brooding, some communication between the sexes might be necessary. This is, of course, true only if both sexes are active in brooding. In many animals, either the male alone, or more often the female alone, takes care of the eggs, and drives the other sex away, treating that individual as an intruder. In these cases, the usual territorial communication signals by the brooding parent are used to ward off invasion of the territory.

Brooding may be a full-time job, or may involve only preparation of a suitable place for the eggs, with little other care. For example, dung beetles roll up balls of dung, to harbor the eggs, the male helping the female. The female then lays an egg in each ball of dung, and often stands by to guard these little masses of food for the developing young. In one beetle, *Geotrupes*, the male and female together dig a hole into which they place the little food parcels for the expected young. The female, in this case, takes on herself the duty of rolling these up and making them into the proper shape, while the male carries the material into the hole. If the male tires of this back and forth duty, leaving the female with no material to work on, she stridulates noisily to call him back to work. Sometimes he does not listen to her "voice," and she then rushes out and uses her mandibles to impress upon him the urgency of the occasion.

Social insects, such as ants and bees, show the greatest complication in communication during brood care. In the case of the honey bee, for instance, the eggs from the queen are deposited in the wax comb, and there the larval bees are fed by the workers. This involves an elaborate communication network. The larval bees must learn to solicit food from the workers, and the workers must learn to recognize the solicitation signals. The eggs of the queen are laid in different sized cells, so that three sizes of bees are produced: worker, queen, and drone. These are given different foods for their different developmental needs. The bees know, from the size of the cells, the nature of the food to be given.

When worker bees emerge as adults, they enter an entirely new world from the protected one they formerly had. They now become nurse bees, and their first job in the hive is to take care of the larval bees, which they themselves were so short a time ago. Later they must ventilate and defend the nest, and identify returning workers. Still later they become foragers, and finally they are scouts, in which capacity they use the elaborate dance rituals. From larval life on, the bees are involved in the use of elaborate communication signals mostly based on antennal contacts and body movements.

Birds take great care of their developing eggs. Either before or after mating, the male and female bird cooperate in building a nest. Nest construction is a complicated behavior pattern and usually involves communication signals between the sexes. Obviously, the usual social signals enter into communication patterns during this nest-preparation period, and, in some birds, courtship involves simulated or real nest construction.

In most birds, both sexes are involved in the incubation of the eggs, and there are various patterns of care. The female or male may remain on the nest, and the other individual act as a feeder. On the other hand, both sexes may take turns in sitting on the eggs and incubating them. In the latter situation, it is necessary that the individuals be able to identify each other in nest-exchange ceremonies. These ceremonies are usually visual and auditory, as one would expect with birds. Often they involve some aspects of feeding or mock-feeding, and usually some of the aspects of courtship.

Although in many birds the time of incubation is relatively short—a few weeks—in other birds it may be rather long. In Laysan and Black-footed Albatrosses of the mid-Pacific, the time required for egg hatching is about sixty-five days. The male and female take turns in incubating the eggs; while one is on the egg the other goes out to sea to feed. The bird on the nest may be left for some time—a few weeks or more. When the feeding bird returns, one would imagine that there would be no problem for it to get on the nest, but this is not so. The returning bird has to go through an elaborate program of identification, using special call-notes and some coercion to get the sitter off. When it succeeds in this, the replaced bird leaves and, strangely enough, also stays away at sea for a long time.

PARENT-YOUNG RELATIONSHIPS

When young animals hatch from eggs, they may be entirely on their own, or they may be cared for by the parents. In the former case, there is scarcely any need for communication systems, but, in the latter, it is obvious that a social unit develops, and the maintenance of a social unit requires some means for the individuals in the unit to communicate with each other.

Insects

It would seem only reasonable that communication between parent and young necessitates their being together. However, an exceptional case has been reported in which the parent communicates with the young, even after the parent is dead. This is found in certain wasps.

The female wasp makes a tunnel in a stem and in it places her eggs, with food for the young, in a series of cells separated by mud walls. When the new wasps have become adults, they must crawl to the outside, without destroying all the other cells on the way. Thus, it is necessary that they know which direction to gnaw. The female wasp indicates this by the shape of the mud walls separating each cell from the next. One surface of the wall is concave and smooth, the other surface convex and rough. The young wasp, therefore, must choose between smooth and rough surfaces, or concave and convex surfaces. It gnaws through the convex and rough, and so gets to the outside without going through cells where others are still developing.

Here, again, one can see the use of two signals, when only one is absolutely necessary. But, in a situation such as this, fraught with possible danger for the future of the species, no chances are taken, and a redundant signal is set up. The amazing thing here is that the egg usually hatches after the female is dead, and the message reaches her offspring long after she is gone.

In honey bees, as we have noted, the workers, which are sterile females and not parents, and the developing young communicate with each other. There is also an interesting communication system set up between the real parent, that is, the queen, and her worker daughters. It has long been known that, if the queen is removed from a hive, the bees shortly become agitated, and take steps to replace the queen by moving a young larva to a larger, queen cell in the comb, and by feeding it on royal jelly. Obviously, the fact that the queen is gone is somehow communicated to the workers in the hive, and this brings about an appropriate reaction.

In a sense, this turns out to be a type of reversed communication—the absence of a communication signal, which otherwise is constantly present, triggers off the response. The queen produces in her mandibular glands a material called queen substance, which is picked up by the workers that tend the queen, and, as they share food with others, is passed along through the colony. Each of the workers gets some of the queen substance by accepting food from other bees. This substance inhibits the formation of the ovaries of the workers, and so is part of the system that makes them workers. If the queen is removed from the hive, the queen substance disappears. With its disappearance, some workers may actually become mature females and develop ovaries. More usually, however, the workers quickly bring through another queen to lay eggs for the hive.

Queen substance has recently been extracted and chemically identified as 9-oxodec-trans-2-enoic acid. By using the pure material, biologists have found that this particular substance is not enough, by itself, to inhibit ovary development in workers. The scent of the queen must also be present, and this is another chemical. The queen substance does not act, as one would expect, as a drug, for it is inactive when fed to bees. It probably acts, as an odor or taste, through the chemical sense-organs.

Queen substance is actually somewhat of a chemical enforcer of territory, in that it represses the development of rivals to the queen in the hive. It never acts completely, however. As the population of workers gets larger, apparently not enough queen substance can be produced to go around, and the bees start raising other queens. When this happens, the old queen senses the development of the new queen, and starts to make sharp, piping sounds recognized easily by the human ear. In a sense, she is notifying the rival that this is her territory. The workers seem to be able to receive the sounds, as vibrations, probably through sense-organs in their legs. At least, the bees around the piping queen stand still. If man vibrates the hive with a sound approximately like that of the queen piping, the bees hear it and become immobile. Other queens answer the piping, if it is broadcast in the hive.

When the old queen pipes, the new queen may try it while still inside the queen cell; she then makes a sound that is much like that of the old queen, but, because it comes from inside a cell, it has a deeper pitch. This is called by beekeepers, qualking or quawking, in imitation of the sound. The old queen may now discover where the new queen is and kill her in the cell, or if the new queen has emerged, engage her in a fight. The fight may result in the death of one of the queens, but often it does not, and the colony splits, each queen keeping part of the workers. Usually the old queen leaves, taking with her the swarm to found a new colony. If the disorder was due to the presence of too little queen substance, the splitting of the colony rectifies that; there is again enough queen substance to go around, and development of queens is repressed.

Birds and mammals

In birds, recognition of the young by the parents involves chiefly visual cues. It may seem to us that many birds, babies or adults, look much alike. This, however, is by no means true. If one observes birds closely for a while, he soon learns that individual birds not only look different, but, even more important, they have individual patterns of behavior which are easy to distinguish. Furthermore, birds use their voices for communication, and there are subtle differences between the voices of different individuals. Thus, it is probable that the baby birds, which are recognized at first because they are in the nest that the adults made, are recognized after they have left the nest by their visual and auditory peculiarities.

In mammals, recognition of the young by the parents seems to be chiefly olfactory. As noted before, man is a rather unusual mammal in that his olfactory sense is rudimentary, whereas his visual sense is highly developed. Most mammals rely heavily on their senses of smell and hearing, and have poor vision, indeed many mammals are entirely nocturnal. It is understandable then that the sense of smell is the chief method by which species and individuals are identified.

The recognition of the parents by the young also involves communication. Among birds, these signals are probably, at first, chiefly auditory, but later are mostly visual. This would be particularly true for birds which are born blind. Many species of birds exhibit the phenomenon called imprinting, in which the young fixes upon the animal that it hears or sees first as representing its species. Thereafter, any object that has this auditory or visual pattern is treated as the same species. Young ducklings, which under ordinary circumstances would see ducks first, and therefore associate themselves with ducks, can, by first seeing species such as chickens or man, attach themselves to these animals as if they were members of their own species. In this case, then, visual communication signals are learned almost immediately by the young. Other birds do not show this almost automatic development of means of identification, but have an inborn recognition of the species, usually visual, which is little affected by early experience.

In mammals, the close association between the young and parents seems to have resulted in an almost indiscriminate adoption of surrounding objects by baby mammals as a parent image. Usually the object need only be soft and warm, and sometimes furry. Thus, a baby monkey will treat, as if it were a mother, a doll made of wire and soft material against which it can cuddle. There is perhaps some form of imprinting, but mammals, with their greater learning ability, may learn to ignore this and become more discriminating.

During feeding and nest care by the parents, birds and mammals use a wide variety of signals. Many of these are like those that adults use with each other. Young birds, for instance, solicit food by peeping sounds, or by begging. The baby Herring Gull solicits food from the parent by pecking at a red spot on the end of the parent's beak. Most babies solicit food by gaping and by special sounds. These food-solicitation sounds may be critical in communication; deafened adult birds of some species cannot bring the young through.

When a group of baby birds is in the nest and well fed, they often keep up a low cheeping, which has been named the pleasure call. Naturally, it is impossible to know whether this reflects any pleasure on the part of the birds. It may represent the juvenile form of signals that have been called all's-well calls, when used by adults. These are rather nondescript acoustical signals, produced by birds that are otherwise not apparently stimulated, and seem to reflect merely a state of well-being.

Usually, in both adults and young, a condition in which there is no danger and sufficient food is not indicated by any special signals; these might attract predators. Instead, a satisfactory state of affairs simply results in no alarm or distress calls. In many mammals, facial expressions, which can be seen only at close range, indicate degrees of tranquility or lack thereof. Many primates, including man, can exhibit a whole range of facial expressions to indicate various levels of pleasure, fear, or pain. These signals are readily perceived by the adults when given by the young, and bring responses.

Obviously, just as the young signal to the parents to maintain the parent-young relationship, so the parents signal to the young. The mere appearance of a parent bird at a nest induces the babies to gape and solicit food. At first, this reaction is quite nonspecific: the baby birds gape and peep at almost any object above them. Later, however, when their visual or auditory senses develop, so that they can distinguish their parents, they gape only toward them and cringe or remain quiet when other individuals come near.

Although young baby birds, by themselves, may gape at all

sorts of outsiders, the adult birds, if they see the invaders, can use the alarm call to signal to the babies. Usually, the first reaction of baby birds to the alarm call is freezing or immobility, which draws them into the cover of the nest. This reaction is particularly marked in young gallinaceous birds, or baby sea birds. These babies are colored almost exactly like the ground on which they live and, if they remain absolutely quiet and hunched together, are almost invisible. Baby Herring Gulls exhibit a fine example of this type of reaction. If a prospective enemy appears, the adult birds in the air shout the alarm call sharply, and the babies cringe against the earth, where they are extremely difficult to see.

The freezing reaction of baby birds to the alarm calls of their species seems to be generally inborn, or to develop early in life. Young Turnstones, in England, have an inborn freezing reaction, and they do not respond to the alarm call of other species. However, if Turnstone eggs are incubated by another bird, the Redshanks, the baby Turnstones respond to the calls of the Redshanks. Thus, while the tendency to freeze at an alarm call is inborn, the specific signal to which it is done is partly learned.

In mammals, the main signals of parents to young, at first, seem to be chiefly mechanical. The solicitous licking of kittens by a mother cat, the nuzzling of puppies by a dog, the cuddling of baby primates by the mothers, all show this mechanical communication. Later, as the eyes of the babies open, visual and auditory signals become more important. In mammals that are born in a relatively independent state, such as cattle, elk, and deer, the young almost immediately respond to adult visual and auditory signals. Their reactions may be different from those of the adults, however. Adult deer, for example, respond to the alarm signals—a tail flash or a snort—by fleeing. The fawns remain totally immobile and as flat to the ground as they can. With their beautifully spotted coats, they match the forest floor, and are inconspicuous. When the fawns are capable of running with the adults, they lose this reaction to the alarm call and develop the adult reaction.

When the visual and auditory senses have developed sufficiently in baby mammals, the young react to the subtle auditory cues and varied facial expressions which the adult mammals produce with an extended repertory of responses. As maturity comes on, the young gradually show adult reactions to all adult communication signals. Finally, at maturity, they take upon themselves responses to all the complicated social and sexual signals of the species.

DEVELOPMENT OF COMMUNICATION PATTERNS

Just as the body grows and the form changes during an animal's development, so its behavior patterns grow and change. As in every other phase of animal communication, these changes depend upon the capabilities of the animal—upon its communication systems at the beginning, and its potential communication channels. Studies on the development of the communication patterns enable us to understand better the nature of adult communication patterns, which are often complex.

Invertebrates

In most invertebrates, the production of communication signals requires the development of special structures, such as the large and showy claws of male fiddler crabs, the special color patterns on the wings of butterflies, or the bushy, hairy patches on the legs of spiders. These develop as the animals mature, and so the immature forms do not have the structures needed to make the appropriate signals. For instance, immature butterflies are caterpillars, and, although they have their own color patterns and may show territorial behavior and warning displays, they do not have wings to produce the sexual signals of the adults.

When an invertebrate larva becomes an adult, it is usually not near older adults, yet it immediately begins to use its special structures or colors in the particular communication system of the species. It is possible for some individuals to learn to vary the communication systems during their lifetime, but it is generally impossible for them to pass this on to future generations. Only if the particular variation is genetically inherited and confers some advantage on the possessor may it be passed on to the next generation. In this way, the genes determining the new pattern may gradually displace their predecessors and effect changes in the behavior patterns of the progeny.

Insects that sing generally stridulate only as adults. There are a few grasshoppers in which the young, the nymphs, produce sounds; this is usually accomplished by rubbing the legs against the body or tapping the body against the ground. These sounds are probably used as simple territorial or warning signals. However, immature grasshoppers and crickets do not have the complicated ears that are found in the adult, so they cannot distinguish the subtle patterns of sound, which the adults can produce.

When the nymph makes its final molt to become an adult, it not only acquires the highly specialized ears on the legs or abdomen, but also develops the mechanism for producing the specific sounds for adult life. Here again, without any practice or direction, the adult insect produces the sounds that are characteristic of the species. This does not mean that every adult insect necessarily produces exactly the same sounds, for there are individual differences, even among these creatures. It does mean, however, that if an individual changes its own singing patterns, without having the change genetically based, the new pattern will not be passed on to the next generation.

Vertebrates

Among the lower vertebrates—fish, amphibians, and reptiles—as with the invertebrates, it is generally only adults that produce most communication signals. It would be interesting to know whether tadpoles have a communication system; recent work on social organization of some groups of spade-foot toad tadpoles has indicated that there is a social structure, which might require some sort of communication. However, studies on this are too few to make general statements. Young fish form schools, and use body patterns or odors for recognition, as we have seen. The full battery of communication signals of fish, however, develops only with sexual maturity. Adult fish show the behavior patterns of the species essentially perfectly as soon as they have acquired the necessary colors, organs, or propensities; no practice or coaching is necessary.

In birds and mammals, where parental care is the rule, and where sense-organs and behavioral capabilities are highly evolved, the development of the communication systems involves feedback from parents to young and probably also from young to parents.

Baby birds have, at hatching, the ability to produce food-solicitation signals, usually some form of all's-well or pleasure signals, and sometimes a few others. They can react at the appearance of, or on contact with, an adult with appropriate responses—begging and gaping for food. They generally have a response to the alarm call—freezing or cringing. They do not respond to the territorial song or to the many other signals that the parents use in sexual communication.

As the young develop, their ability to produce the various social facilitation calls develops too. As we have seen, young birds usually learn to identify their parents by having the shape imprinted upon them early in life. They gradually develop the ability, through seeing other members of their species, to distinguish the species. Baby birds that are raised in the nests of other species may adopt the other species and stay with them. Even in these cases, however, as maturity comes on, the ability to distinguish the true species seems to override the learned mistake.

When young birds become mature and are ready to breed, they begin to sing with the characteristic territorial and attractive songs. In some birds, such as blackbirds, the complex song pattern seems to be pretty well inborn: a young male that has never heard any other bird sing produces the song just as does an older male. In other species, such as the Chaffinch, a young male that has never heard another male produces only a skeleton of the species song. Only if it hears singing males of its own species, does it develop the full song. This is probably true also of American thrushes, which use a number of themes and recombine them to form many different songs. In these cases, the birds imitate the themes of other birds around them, and, by adding these to their own repertory, produce amazingly complex song patterns.

In general, the reactions of male and female birds to sexual signals seem to be mostly inborn. Many of the sexual displays are derived from those associated with immature behavior. In many birds, the female solicits food from the male, and the male indulges in feeding, or ritualized feeding, of the female. Thus, sexual behavior may be influenced by individual or even racial experience, for young birds may copy the actions of adults.

In mammals, much the same situation exists. The babies have only a few responses at birth, and these are usually quite stereotyped. As their sensory capabilities emerge, they develop the ability to distinguish their parents from other individuals and to communicate more exactly with them. Because of the much greater variability in behavior patterns and communication signals, young mammals learn a great many more varied responses than the young of any other group of animals.

As with most other animals, however, reactions to sexual signals are mostly inborn, and, when the animals become mature, they produce and react to the signals as their parents did, generally without any training. Where young mammals stay together in a family, as in man and other primates, however, the young indulge in sexual play before there is any sexual competence. Without this practice, the complicated courtship and mating patterns used by primates apparently develop most unsatisfactorily.

Types of developmental patterns

Most animals produce signals and react to those signals only at certain times in the life cycle. Some communication signals evoke strongly stereotyped behavior; in lower animals, almost all do. In higher animals more and more learning occurs as the young mature, and more and more signals are built into complex social systems.

In lower animals, with little social life, individually learned differences in behavior patterns cannot be transmitted to the young. Only genetically determined ability or lack of ability to produce variations in communication behavior can be passed on. If this ability enables an animal to succeed better in the race for reproduction, that particular feature is favored in the competition and is more likely to be handed on to the next generation. Given continuing success, the new behavior pattern can ultimately be incorporated into the species.

In higher animals, such as social insects, birds, and mammals, behavior patterns can be changed much more drastically within the life of the individual. Only minor variations, however, can be incorporated, by learning of offspring, into the species. It is only in man, who can produce communication signals lasting beyond single lives, that changes in basic behavior patterns can be transmitted from generation to generation without being incorporated into the genetic code of the reproductive cells.

9

Efficiency and Evolutionary Importance

MANY times, up to this point, we have mentioned that communication—particularly that involved in sexual reproduction—is important in the evolution of animals. We have suggested, in other words, that the reproductive communication signals of animals may be significant factors in maintenance of species purity, and possibly also in the emergence of new species. It is worthwhile, at this point, to examine this idea in more detail.

However, before we do that, we should discuss one other facet of animal communication—the efficiency with which it operates.

Theoretically, at least, information should be transmitted by a communication system with minimum risk of error. This means that the signals should be specific and distinct. They should come through clearly above the noise level created by the presence of other energy or chemicals in the same channels. In the case of social cooperation signals, there might be little loss of value, if the systems were not species-specific, provided useless random responses did not waste too much time. In the case of reproductive signals, however, it is important that the information be species-specific, and that little or no chance of error exist.

SOURCES OF ERROR IN ANIMAL COMMUNICATION

Perfection in a communication system implies that the information be transmitted unequivocally, so that random responses are avoided. There are two main deviations from this perfection found in animals: first, the development of dialects—if we may call them that—in which some individuals, because of peculiarities in their signals, are unable to communicate with others of the same species; second, cross-reactivity, in which the signals of one species bring about reactions in another species as well.

Dialects

Dialects in communication systems have been discovered for a number of animals, even though exact studies of them have been few. The reported cases are found chiefly within the three groups of animals that make the greatest use of communication signals: birds, mammals, and insects.

Many field ornithologists can give examples, from their experience, of geographic differences in songs or calls of birds of the same species. Now that the tape recorder allows us to collect songs for comparison and physical analysis, the differences in songs among geographic races of birds can be studied more critically.

In some birds, the territorial song is apparently well fixed at the time of hatching. In these species, little variation in song pattern occurs, even when groups exist in widely separated geographic areas. However, in many birds, only the skeleton of the song is inborn, and most of it is learned by imitation. In these cases, differences in songs in different geographic areas can result in dialects.

In White-crowned Sparrows, Marler has reported dialects in three separated geographic areas of California. The farther removed the individuals are geographically from each other, the greater the differences. A similar situation exists with respect to the Apapane, an Hawaiian bird. Dialects have been found, as might possibly be expected, on different islands of the Hawaiian chain. On the island of Kauai, however, geographically isolated groups, as close together as six miles, sing with slightly different songs.

The complex songs of birds offer a rich opportunity for variation. Even the call notes, however, which are generally simple in nature, may vary from place to place. For example, the calls of Herring Gulls recorded in Maine are unintelligible to birds of the same species in Holland and France. It is interesting to note that, on the bases of other features, it has been assumed that the Herring Gull originated near Alaska and spread from the point of origin in both directions around the globe. On this basis, therefore, the two groups separated by the Atlantic Ocean would be the most distinct, in terms of sharing of genetic material.

In correspondence with the development of different signals within the same species, the responses to the same signal may vary within the species. Thus, some individuals develop broad responses to signals and can understand the calls not only of members of their own immediate geographic group, but also those of other groups. On the other hand, some individuals become limited in response to the calls of only those members of the species found in a given area. Some birds learn to respond to the calls of a number of races of the species or even to the calls of other species by fraternizing with them, whereas others become extremely provincial in their responses.

This sort of situation seems to exist in the case of Eastern Crows in Maine. They meet only Crows from the same region, and respond only to their specific calls. On the other hand, Eastern Crows in the more south-easterly states, which move up and down the coast from summer to winter, and so meet a related species, the Fish Crow, respond to a wider variety of calls.

Even in cases in which the communication system is not vocal, and the word dialect might not seem appropriate, there are differences in communication systems from race to race.

The German race of honey bees performs the round dance, when food sources are closer than about 275 feet from the hive. Beyond this, the bees perform the waggle dance to indicate direction and distance. On the other hand, the Italian race of the same species uses the round dance only when food is closer than about 30 feet from the hive. If the food is between 30 and 120 feet away, the Italian bees use the sickle dance, which is different from any used by the German bees. For food that is farther than 120 feet, the Italian bees use the waggle dance. However, in indicating the distance in the waggle dance, the Italian bee turns more slowly than the German bee. If the food is about 1000 feet from the hive, for example, the German bee performs about thirty waggle-runs per minute; the Italian bee performs about twenty-six waggle-runs per minute.

As noted before, honey bees of one race will allow members of other races of the same species to live with them, if the foreign bees are placed for a few days in a protective cage in the hive to acquire the hive odor. The immigrant bees, however, when giving guidance signals misdirect their hosts as to distance, because of their dialect. The German worker bees fly too far, when given directions by members of the Italian race. Certainly this closely parallels what happens in the human race when misunderstandings occur in subtle meanings of words.

In the United States, there are species of crickets in which males of different races use different sounds for calling the females. In *Anaxipha exigua*, for instance, three differently singing races have been discovered. On the basis of structure they cannot be distinguished, so they have not been considered as true species. One of these is a woods-inhabiting form; it has a chirping song. The other two live together in fields; they have trilling songs, one a fast trill, the other a slow trill. These three can be mated with each other, but usually in nature their songs prevent this, for each female responds only to one of the three types of song.

The colors and designs on butterflies' wings are used as mating signals, and here too there is evidence that dialects—if we could use such a term—exist. The wood nymph butterfly, *Cercyonis pegala*, of the Eastern United States is characterized, in its southern range, by clearly marked eye-spots on the wings, in its northern range, by indistinct eye-spots; in between, intermediate forms and mixtures occur. It is quite probable, although studies are lacking, that we would find differences in the reactions of females at different points along the continuum of forms from one end of the Allegheny Mountains to the other.

The development of dialects within species may be important biologically, for this may start the separation of a population into sub-populations, whose reactions to communication signals gradually become so different that they no longer share the same genetic background. When this happens, they are essentially two species, for it is only a matter of time until chance, but different, genetic changes will be fixed in the groups and thus separate them structurally and finally in ability to produce fertile eggs.

Cross-reactivity

The second deviation from perfection in communication systems is the development of cross-reactivity. This means the development of reactions by a receiver to signals other than those that are appropriate to the species for the time and place. In many cases, this may have little effect, except to introduce higher noise-level. If it happens in sexual behavior, and other signals are lacking, so that there would be attempts at cross-mating, it might be regarded as biologically unprofitable. If, on the other hand, it occurs between two species that are sharing food sources or common enemies, it could result in interspecific cooperation.

Cross-reactivity is best seen, experimentally, in reactions of animals to artificial signals. Often, when captive or wild animals are subjected to artificial signals, their reactions do not obviously resemble the normal reactions to communication signals. One must, of course, always be wary that the signal is like or unlike that of the animal, in the animal's terms. Human beings, being unable to hear ultrasonic sound or see ultraviolet light, may think that these take no part in animal communication systems. We know, however, that there are some insects which can receive these particular forms of energy, and use them for communication.

In an attempt to discover the essential features of the communication signals of some grasshoppers in France, Busnel played back to them recordings of the communication signals of their own and other species, and bird calls or pure tones as well. The reactions of the insects seemed to be rather unspecific at times. They often answered sounds that had little resemblance to their own sounds. Busnel believes that the common feature in the effective sound signals is the presence of some sudden start or stop, a transient, and that transients, in themselves, induce reactions in many insects.

This may, however, not be the case. We have found reactions by some insects to sounds that increased in amplitude gradually, and therefore did not have any transients. The feature eliciting reactions to acoustical signals in insects seems more likely to be the rhythm of the signals—the number of pulses, and their timing. Much more work will be needed before the matter is settled, but it would look as if responses to artificial sounds are unspecific reactions to signals that somehow partake of pulse characteristics of the insects' signals themselves.

More amazing, however, Busnel found, in testing the singing of male grasshoppers of two different species of *Ephippiger*, using the female of one as the reactor, that the female was attracted just as often to a male not of her species as to the correct male. This indicates a communication system which is inefficient. However, in nature, as Busnel points out, these two species do not live near each other; they had been brought together for this experiment.

This places the whole experiment within a broader framework. Blair, in the United States, has found, for frogs, that, if two species are similar in size and appearance and have overlapping ranges—therefore, could hybridize—their songs are quite distinct. If the two species do not have overlapping ranges, but are separated by a considerable distance, their songs may be similar.

If a species has a particular type of sound-producing equipment, there are probably only so many possible variations. Consequently, if two species live widely separated geographically, or breed at different seasons of the year, specific differences in the communication signals do not develop. On the other hand, if the two species breed in the same region at the same time, so that the females could be attracted to the wrong males and thus waste their reproductive cells, specific signals of males and specific reactions of females to the signals evolve.

Now, going back to the grasshopper experiment, the same sort of thing could be the case here. The related species of *Ephippiger* probably have similar songs, but they live in different regions; thus there is no chance of confusion. It comes down to a matter of how broad a spectrum of response is permitted a receiver, within any given environment, before it starts responding to incorrect signals, thus wasting energy. If a species of singing insect lives in an acoustically simple environment, the females can respond to almost any sound, and they usually find the right male. On the other hand, if the species lives in an acoustically complex environment, the reactions of the females, if they are to be biologically useful, must be more sophisticated.

So far we have discussed the development of cross-reactivity as if it were disadvantageous. However, it is only in sexual signaling, or in cases in which the animal becomes responsive to almost any stimulus in the environment, that cross-reactivity is unequivocally harmful. In cases in which species learn to react appropriately to social facilitation signals of other species, a form of cooperation may develop. Obviously, this originates by learning, for it seems unlikely that animals would have reactions to communication systems of other species as part of their inborn behavior.

It is a common observation that alarm calls or mobbing calls of many birds excite reactions not only in the species producing the call, but in other species as well. If small birds set up their mobbing call around an owl, they usually attract birds of many other species, which join in the call. Likewise, many species of birds respond to the alarm calls of other species in their vicinity by appropriately flying away.

An interesting case of this occurs in the Eastern Crow and Herring Gull in Maine. On Mount Desert Island, where the Gulls and Crows rarely feed in the same spot, the two act almost independently of each other's signals. However, only a few miles away, on Schoodic Peninsula, where tourists feed the Gulls and, inadvertently, the Crows too, the birds respond to the food and alarm calls of both indiscriminately. In other words, where the two feed and travel together, they have become, in their communication systems, part of one family, so to speak—a form of cooperation that is advantageous to both.

This type of cooperation may even involve animals of different classes. For example, in Africa, the rhinoceros responds to the alarm calls of the so-called Rhino Birds, which walk about on the mammals and feed upon their ectoparasites. In other places, Cow Birds or Cattle Egrets, which frequent places where there are large ungulates, warn the mammals of approaching enemies by their alarm calls.

Similarly, where animals of different groups try to use the same territories for feeding or breeding, an understanding of the defense signals may develop interspecifically. Some species of mammals or birds, for instance, defend their territories not only against other members of the same species, but also against members of other species. Usually the contests for space do not come to true conflict, but are resolved by territorial displays, understood mutually.

These last examples suggest the possibility that learning may not be involved in all cases of interspecific cross-reactivity. An alternative explanation would be that there are essential similarities in alarm signals and territorial defense rituals of many species. Interestingly enough, even untrained human beings may find it easy to recognize defense and threat postures in mammals and birds. Thus, growling and baring of teeth by mammals, when they are threatened or are threatening, are understood by human beings without any particular learning being needed. This is a common type of threat behavior in all mammals, even man.

Marler has suggested that there is a basic similarity in the alarm calls of many species of song birds. He has found that the hawk alarm calls and ground alarm calls of many are similar. Thus, birds need not learn the alarm calls of other species to respond to them. They may respond to basic features of the signals which are alike for many species. However, learning would seem to be involved in the case of the crows and gulls cited above, and we can not rule out this possibility in any of the cases mentioned. It is extremely difficult to distinguish learned behavior from innate behavior.

As noted, man has little trouble understanding the threat postures or distress calls of many animals. Something in the sounds apparently comes through at a deeper than language level. We all know that sobbing, crying, or moaning—regardless of the language of the individual—communicate distress. There may be levels of communication, involving all our senses, which are deeply rooted in pre-human origins, lying below the level of immediately perceivable gestures and language. This aspect of human communication still needs pioneering study.

ANIMAL COMMUNICATION AND EVOLUTION

Evolution is defined as descent with modification. Over the millions of years of the earth's history, animals and plants have changed, adapting themselves to changing environments, and generally increasing in size and complexity, to produce the vast array of organisms that are alive today, all remarkably suited to live where they do.

All modern biologists believe that evolution of living creatures has occurred. However, they are by no means agreed as to how it came about, and one of the most active fields of study and thought in biology today is evolutionary theory. We shall very sketchily outline the processes involved in evolution, as conceived by modern biologists.

The basic unit which evolves is not the individual, but the population. Population, as used here, refers to any group of animals or plants the members of which are similar in appearance and freely exchange, within the sexual framework, their heritable materials. Populations may range in size from small groups restricted to tiny areas—under a single stone, or in a puddle of water—to large groups extending over vast areas. Small populations are usually called demes; large groups may represent all the members of a wide-ranging species.

The heritable units of a population are the genes, and the sum total of genes in a population is its gene pool. As each individual is produced, it, figuratively speaking, reaches—through its two parents—into the gene pool of the population and draws out a sample of genes, which interact with the environment to determine the exact individual plant or animal it will be. Since the genes are drawn by each individual at random from the gene pool and many individuals are usually produced, every generation has almost exactly the same proportion of genes as the previous generation, even though the individuals vary tremendously.

The frequencies of genes remain essentially constant through succeeding generations, however, only if: no new genes are introduced, either by mutation or introduction; reproduction is random, so that every gene gets an equal chance of being passed on; and all the individuals expressing combinations of genes have an equal chance to survive. In natural populations, as it so happens, one or all of the conditions necessary for stability of the gene pool are violated.

Changes in chromosomes or genes, mutations, are constantly occurring. The new mutant genes are added to the old gene pool. With continued mutation pressure, if no removal of new genes occurs, the frequencies of these genes gradually increase. Ordinarily new genes build up in a population until they themselves start to mutate at a rate equal to the rate at which they are produced, thus establishing an equilibrium. In natural populations, most genes, but not all by far, have reached mutation equilibrium.

The gene pool is also affected by nonrandom breeding. Nonrandom breeding can be brought about by environmental means, or by changes within the animals or plants themselves.

Two environmental factors that result in nonrandom breeding are population reduction and geographic isolation. If a population falls to a very low level, then the chances for certain genes to be lost, merely because there are not enough individuals to get all of them, become increasingly high. This is known as genetic drift, and accounts for loss of characters, which accidentally are not passed on to the next generation. Geographic isolation also may result in nonrandom breeding. If a population of some species becomes isolated on an island, let us say, it brings with it only the genes that it has. As a consequence, the isolated animals or plants will form a new population, with restricted gene pool at first, and later with added mutants differing from those occurring in the parent group.

Physiological factors that result in nonrandom mating cause two types of isolation that are of great importance. First, if different individuals in a species reproduce at different seasons of the year, they become isolated from each other, just as if they were isolated geographically. This is called temporal isolation. For instance, the field cricket, in certain regions of the United States, has two races: the individuals of one race mature in the spring, those of the other mature in the fall. The spring adults are gone by the time the fall adults mature. It is, therefore, impossible for these two populations to share their gene pools; they are just as isolated as if they were on separate islands.

The second type of physiological isolation is behavioral. In this case, two populations of a species become so different in behavior that they no longer recognize each other as of the same species. There are, for example, two groups in the same species of tree cricket, *Oecanthus niveus*, with different behavior and song patterns. These two have become quite distinct, because the females of one group do not respond to the song of the males of the other group, and so do not mate with them. Thus the two populations no longer share their genes and either are now or will probably soon be separate species.

Sexual selection might be regarded as a special case of behavioral isolation. In this case, females of the population select males with special patterns, or males capable of producing certain communication signals, in preference to others. This produces nonrandom breeding, and tends to favor the genes carried by the males with the striking patterns.

The third criterion for genetic stability of a population is that all gene combinations produce individuals that survive equally well. This is rarely, if ever, true, for the varied gene combinations in a species produce a wide variety of individuals, which are then forced to compete with each other and with the environment to survive and reproduce. Those individuals that have characteristics enabling them to survive or reproduce better are favored to pass on their genes to their offspring. This is Darwin's natural selection.

Natural selection, unlike genetic drift or isolation, is a creative process, because selection for ability to live and reproduce ultimately produces animals best equipped to do just that—they are adapted. Genes that do not contribute to survival and reproductive fitness are gradually removed, and the genetic equilibrium is directionally disturbed.

Natural selection, therefore, gradually produces—within a species—special populations adapted to live in special habitats, thus perfecting the species itself. It does not result in the production of new species. Yet the production of new species—speciation—is the crux of evolution. How, then, can speciation occur?

This is a matter of considerable debate. It is generally admitted that we may designate two populations as separate species when they no longer can successfully interbreed. This does not mean that they no longer do interbreed, just because, for instance, they are geographically separated. It means that they are incapable of interbreeding to produce fertile offspring, even if brought together. The key, then, to speciation is the origin of the barrier against interbreeding.

Many workers feel that, as long as organisms of the same species are near each other and can potentially exchange genes, they will do so. Therefore, they believe, the only way a genetic barrier against interbreeding can arise is for two populations to become geographically separated. If they are separated long enough, they develop genetic differences, through different mutations. Ultimately the differences become large enough and the animals or plants can no longer successfully exchange genetic material; any offspring produced are sterile. If the two populations are brought together, at this stage, there will be no possibility of sharing gene pools. The biologists who believe this, therefore, consider geographic isolation the only acceptable avenue for speciation. Other biologists, however, believe that any type of isolation, provided it is secure enough, can allow two populations, even though living together, to acquire differences in their gene pools great enough that they no longer can share them. Only further research will show which view is correct.

Communication and evolution

Obviously, communication is important in some of the aspects of evolution that we have mentioned. There is a great deal of evidence to indicate that interspecific matings are reduced by specific communication signals. Hybridization between species is very rare in nature; this is true even with species that readily hybridize under artificial conditions. In zoos, numerous species of mammals cross-mate to produce hybrids which are often fertile. However, these same animals, living in the same areas in the wild, do not cross-mate. It is obvious that they do not, because, in the wild, with members of their own species available, they select their own.

Animals of different species may be so similar in appearance that they are almost indistinguishable, but they are usually quite distinct in the olfactory or acoustical signals they produce. For instance, five species of tree crickets, *Oecanthus*, of the United States are almost identical in appearance. Yet their songs, which are the reproductive signals by which the females find the males, are extremely distinct, even to man.

Furthermore, the songs of such similar appearing species are much more distinctive if the animals live in the same region than if they are separated. For instance, two species of tree frogs, which are geographically partly separated and partly together, have distinctly different songs where their ranges overlap, but similar songs where they are separated.

The divergences in signals between species are usually greater in males than in females. Biologically this makes sense, for males produce tremendous quantities of sperms and can waste them with little real energy loss. However, females—of higher animals at least—produce complicated, energy-rich, eggs and lose much more energy if these are wasted. It is critical that a female select the right male. Therefore, selection, in most of the higher animals, falls to the female, assuring that her valuable reproductive cells are not wasted.

In sexual selection, a female bird, for instance, selects the most distinct, the loudest, the brightest male. His brightness and loudness, however, reduce his ability to escape from predators. As a consequence, the male becomes brighter and noisier only to a certain limit. The pressure of predators attacking him, because he is too conspicuous, favors his being less conspicuous. Thus, two selective pressures work on male birds to produce those that are bright enough to be clearly marked for the female, but not so bright that they become easy to catch.

Specific identifying characters often disappear, if species are separated from their near relatives. For instance, two species living geographically together usually have distinct signals. If a population of one is transferred to an island, where the other species does not exist, the group on the island gradually loses the specificity of its signals. All these facts, therefore, lead us to believe that differences in communication signals used in sexual attraction and courtship are important deterrents to interbreeding between species.

In respect to natural selection, there are many ways in which communication signals may be considered to be adaptations enabling individuals to live and pass on their genes to another generation. We shall briefly mention only a few.

Territorial and flocking behavior, enforced or aided by communication signals, certainly help in the survival of individuals through food sharing and cooperative behavior. Food calls and alarm calls enable animals to take advantage of rich food sources and to escape from enemies. Young animals that respond appropriately to communication signals of their parents and other adults of the species have a greater chance for survival than young that fail to respond, and the more exact their response, the greater their chances.

In view of the fact that differences in reproductive communication signals can reduce mating between populations, it seems probable that speciation could be brought about by changes in communication signals, thus not necessitating geographic separation. One might imagine it happening as follows. In a population of insects, some males sing faster than others, the differences probably being genetically determined. Some females, likewise, respond preferentially to higher speeds and some to lower speeds, the differences again being genetically determined. Since individuals producing or responding to higher speeds probably carry many genes determining these responses, their offspring would almost certainly produce and react to the higher speed signals. The converse would also be true.

When the two populations become so distinct in speed and reaction that they no longer respond to the signals of the other group, they would be behaviorally isolated. At this point, they might still be fully capable of having fertile offspring with each other, but, in nature, this would not occur, simply because the females of the one group would have nothing to do with the males of the other.

When behavioral isolation has thus occurred, the populations would no longer be sharing the same gene pool. Therefore, mutations arising in one population would not be passed on to the other. If the mutations in one of these populations had particular survival value, these mutations would build up in that population, but would not be present in the other one. As with geographic isolation, if the gene pools were separated long enough for chromosomal or genetic changes to become distinctive, a genetic barrier to interbreeding would arise. At this point, we would have genetic isolation, and no one would doubt that the species were different. Differences in communication signals and responses may, therefore, initiate the series of changes needed for the basic evolutionary process.

Evolution of communication

Obviously communication signals themselves evolve, just as do all activities and structures of animals. In general, the study of animal communication is still in the pioneering stages, and so we know little about this evolution. Behavior does not fossilize; thus we do not have records, preserved for the ages, as we have with bones or teeth, to give us some ideas of the past. We can only interpret what we see in the present—and we know far too little.

From what we can see, it seems that communication signals usually originate as modified intention movements. An animal makes a movement as if to do something—an intention movement—and this movement gradually becomes a signal to its fellows. This type of communication signal is said to be iconic, for it is an image of the intended act.

An example is found in the stridulation of the short-horned grasshoppers. Almost all grasshoppers move the hind legs up and down alongside the body before copulation. If the legs rub against the wing covers, a light, rustling sound is produced. This sound probably adds to the attractiveness of the movements, and selection by females would be for males producing the sound. Males possessing a structure to increase the intensity of sound they produce would be still further favored. Such a structure would be a thickening of a vein of the wing cover. Thus, males with the stridulating ridge would attract more females and pass on their genes more readily than those without. Gradually, therefore, only males with the ridge, and characteristic leg motions, would be left.

In the honey bee, the communication system may also have originated as a series of complex intention movements. The scout bees, at first, probably brought back the odors of nectar-containing flowers; they could not avoid doing this. They may also have used marking behavior or laying down of odor trails. Later, the bees indicated the course for the others to take by running in the sun, pointing the runs in the direction with respect to the sun that the others should take. This iconic set of directions could then be converted—by what is called ritualization—into the dances that the honey bee performs in the darkness of the hive.

In the courtship of birds, the evolution of iconic to ritualized signals seems to be generally admitted. Most of the courtship patterns of birds include features from behavior during nest building, egg laying, feeding of the young, and so on. In primitive displays, direct nest building by the male and leading of the female to the nest are the central features. Later, these become ritualized: the males build elaborate structures, such as bowers or courting areas, into which they lead the females.

The most highly evolved communication system of all is that of man. It seems to be almost qualitatively different from that of other animals. But is it?

Unfortunately, this question cannot be answered unequivocally at this time, chiefly because we know so little about communication in both man and animals. There is still much debate, for instance, among students of human communication, on the nature of language. Until this is decided, and we know also what we mean by animal language, attempts to compare or contrast human and animal communication are the sheerest speculation.

The chief difference, it would seem, between animal and human communication is that man's language is conceptual and deals with visual symbols. In animals there is little or none of this. Animal languages are more directly related to real objects. Their vocabularies—to use a human concept—consist chiefly of verbs, and these are in the present tense, with implied future. Generally they are imperatives—do this! don't do that! come! go!

The verbs of animals lack the third person and the past tense. When a scout honey bee returns to the hive to signal the location of a food source, she changes her direction as the sun moves. Thus, the honey bee always signals in the present tense. In human language, a person could indicate what he had done, and when he did it, and another person could understand the past tense.

A statement by the British scientist, Haldane, while undoubtedly an oversimplification, illustrates this fundamental difference between animal and human language. He writes that, as long as a child says: "I want this," or "I'm going to do that," it is communicating like an animal; but when the child says, "Yesterday, I did this," it has begun to communicate like a man.

It is the projection of communication into the past and the future that distinguishes man from lower animals. Otherwise, he and they share the same communication patterns, potentialities, and problems.

10

Practical Aspects

It is quite natural that biologists, interested in all phases of animal and plant life, should study animal communication and behavior. This is based on curiosity: the biologists want to know; they care not whether the knowledge has practical utility. The prizing of knowledge for its own sake is a hallmark of civilization, a sign that man has risen above the level of the animal. Throughout history, man has been interested in learning about all aspects of the world around him, even though he cannot turn all of his knowledge to utilitarian ends.

Scientists, however, are willing, if they can, to use their knowledge for man's benefit. They may often feel that they have to emphasize, for persons who fail to see the importance of their fundamental work, the relationship of this work to human development. At the same time, however, they realize that almost any information may influence man's economic and social activities.

And so it is with studies on animal communication. Although the exploitation of animal communication mechanisms for practical purposes is only in its infancy—if even that—we can dimly see some of the many practical values that should accrue in the future. Obviously, since man is an animal, any discussion of animal communication should include his communication systems too. Knowledge of these is of obvious practicality. But what about, let us say, knowledge of the communication systems of roaches, or bees, or minnows, or albatrosses?

IMPROVED HUMAN COMMUNICATION

First of all, the study of communication systems of lower animals should help man to understand his own systems better. In general, human beings favor the auditory channel for immediate communication and the optical channel for delayed communication. It is easy to forget how often we also use other channels—tactile, or possibly even olfactory. Studies on animals that utilize these channels for communication may lead us to understand how we too can use them.

Animal communication mechanisms give us the chance to see the principles of communication in a simpler setting than we see them in man. If we want to do chemical analyses, we try, if possible, to separate the materials into their simplest components and to analyze these individually. If, as in the case of many biological materials, the mixtures become almost hopelessly complicated, our ability to determine exact relationships diminishes. The noise level, if you will, has increased to the point that the signal-to-noise ratio is too low for accurate discrimination. So, if we study only the communication patterns of man, with all their complex interrelationships, we may not see their fundamental bases. Studies on the simpler systems of lower animals may allow us to distinguish clearly the basic mechanisms.

To use any channel for communication means that we must be able to code the transmitted information in this channel. Man has excellent coding systems in the visual and auditory channels, but only rudimentary coding systems in other sensory channels. The use of the Braille alphabet by the blind, transforming visual signals for sighted people into tactile signals for those not so fortunate, is a good example. It may be that our studies on the communicative systems by tactile animals, such as bees or spiders, may suggest other, more efficient, methods for coding this information. The Braille alphabet, obviously, transforms indirectly from one sense to the other. It would be far better to code the information directly into tactile signals.

As man moves into space, new communicative methods may be necessary. It is illuminating that science fiction stories usually have human beings talking to each other, when they are on the moon. With no atmosphere, however, the moon is not suited for acoustical communication, as we know it on earth. If other extraterrestrial objects were densely covered with fog or clouds, visual communication could be equally bad. We might, therefore, want to know whether other sorts of communication signals are possible, and how information can be coded in them.

But we need not turn from the earth to the distant reaches of space to realize that an understanding of communication channels other than the visual and auditory might be useful for man. Knowledge such as this could aid our blind and deaf, particularly those lacking both senses, to find richer lives.

These, however, are not the only persons who might be aided. In some cases, persons who have functional systems for visual and auditory communication do not, or cannot, use them. Thus, psychiatrists find that one factor separating the psychotic or neurotic from his fellow man may be his inability to communicate in accepted channels. This individual may form codes for visual and auditory signals which do not match those of other people. It is as if a honey bee were to try to communicate the location of its food sources to other honey bees in Braille.

The mentally disturbed person may be cut off from normal life simply because he has, in a sense, his own communicative world. However, we might reach these people through other communication channels. Some of these, such as the sense of touch, may be more primitive in a biological sense, so that a much simpler coding system is possible. By using such a system, it might be possible to break through the communication barrier and reach these people again.

BIONICS—SCIENCE OF SENSITIVE SYSTEMS

A second possible use for knowledge of communication systems of animals has just recently been appreciated. A word has been coined to designate this particular use—bionics. Bionics is the attempt to match the actions of sensitive living systems by non-living systems.

Up to now, the imitations, when done at all, have been mechanical or electrical. A microphone, for example, is analogous to the human ear—a vibrating membrane attached to a system converting mechanical into electrical pulses. However, it lacks the exquisite delicacy of the human ear. Furthermore, because in the microphone the living, biophysical-chemical system of the ear has been transposed into a mechanical system, the size is completely wrong—the microphone is enormously larger than the ear. Yet the human ear is hundreds of times larger, and less sensitive to some sounds, than the ear on the front leg of a long-horned grasshopper (Figures 21, 22, and 23).

The conversion of living systems into mechanical and electrical analogues is like the conversion, in Braille, of an efficient written system into an inefficient tactile system. It would be much better if we could duplicate the biochemical and biophysical mechanisms of sensitive living systems. The single receptor cells, even in the human ear, are much smaller than any device man now has for the generation of electrical impulses from mechanical inputs. The receptors of insects called scolopidia are so sensitive to movement that they have essentially reached the ultimate practical limit. Yet these cells are so tiny that a dozen of them can be stretched across the knee of a honey bee. The same could be said of light receptors. The sensitive cells, the rods and cones, of the retina of vertebrates are so sensitive that only one quantum of light is necessary to activate one of them. Yet this bioelectric activity occurs in a far smaller space than that in photoelectric cells.

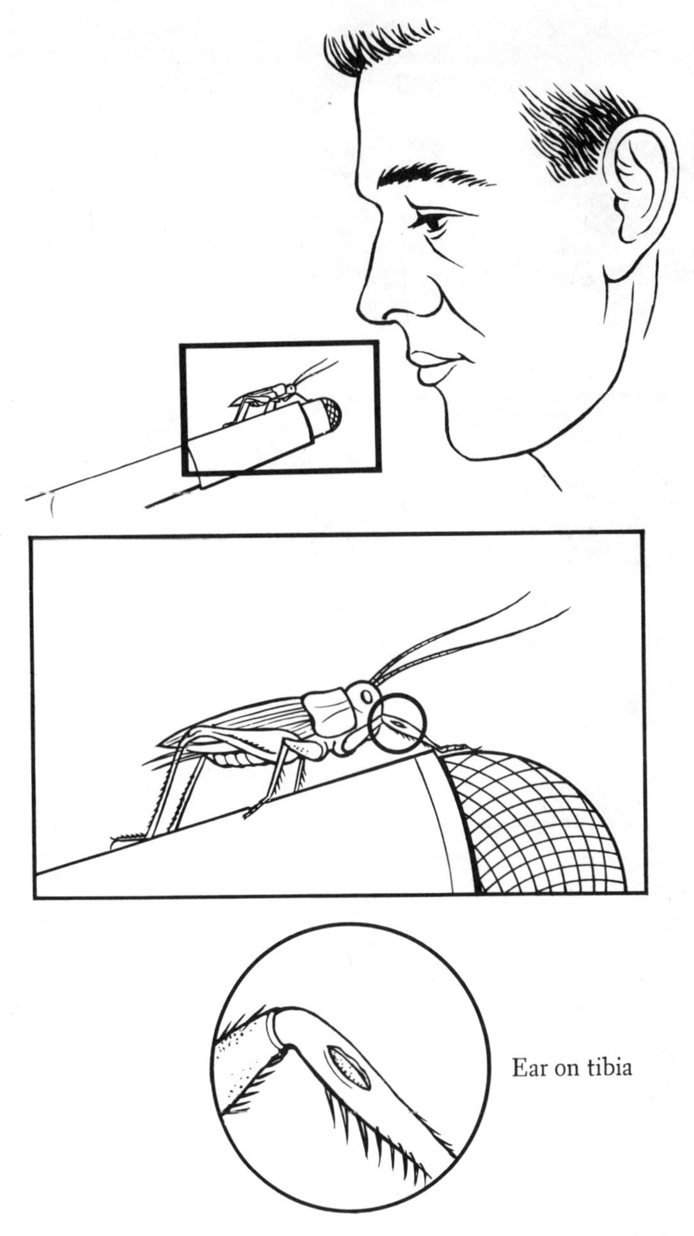

FIGURE 21. *Comparison of size of human microphone and cricket's ear.*

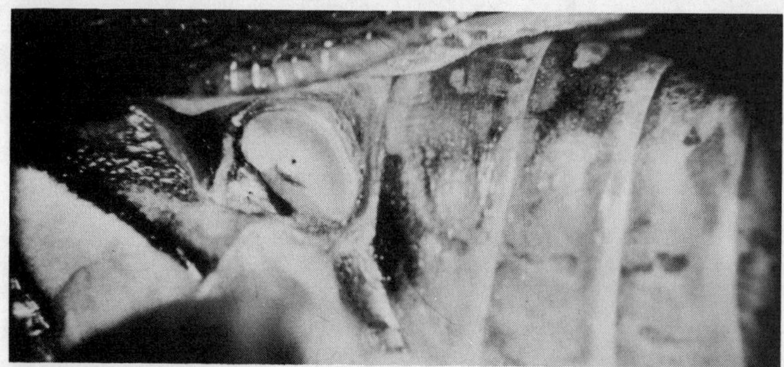

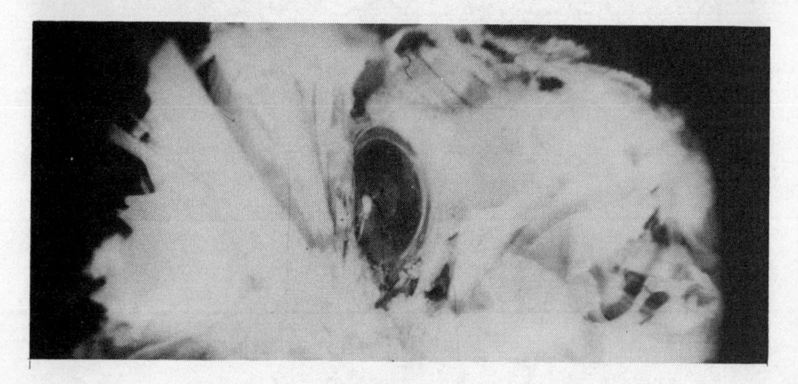

FIGURE 22. Upper: *Short-horn grasshopper (about twice natural size) showing ear near base of jumping leg.* Middle: *Ear (about 8X natural size); note central dot and dark line near it representing a folded area.* Lower: *Internal structure of ear (about 8X natural size); showing nerve attached to ear-drum near center and sensory ganglion near edge.*

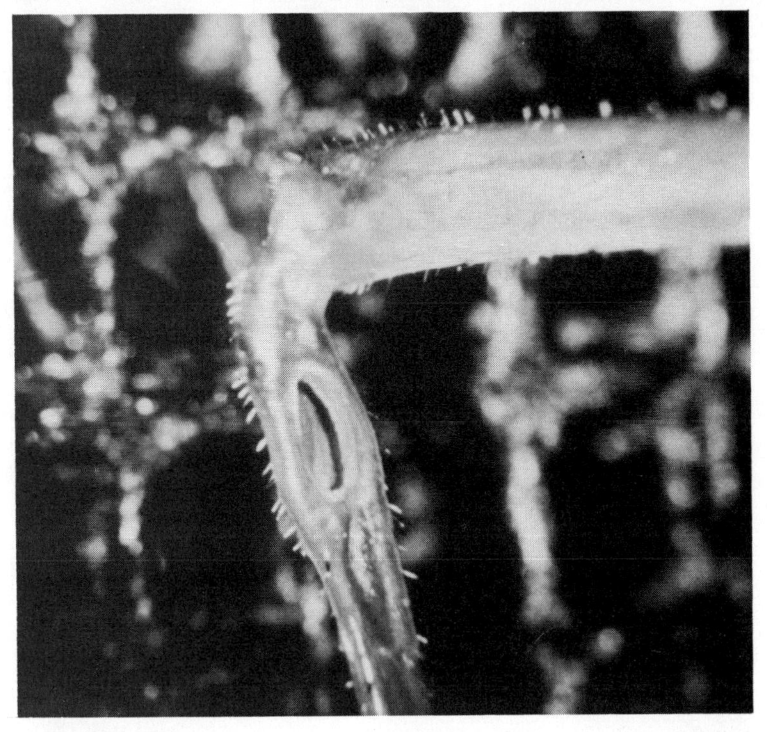

FIGURE 23. *Ear of long-horn grasshopper (about 20X natural size) on front leg.*

If we understood the fundamental biophysical and biochemical mechanisms on which these fantastically sensitive strain and light sensors operate, we might be able to discard the crude materials, such as metal diaphragms and coils of wire, that we now use to create electrical charges. If we knew how the scolopidia of insects generated electrical discharges when they are stretched, we could perhaps use these methods ourselves.

By attaching electrodes to the nerve leading from the ear of a moth, Roeder and Treat were able to pick up, through the ear, ultrasonic pulses produced by bats up to 100 feet away. Human devices are generally not designed to pick up ultrasonic pulses, for man does not hear them. The ear of the moth, therefore, minute as it is, is a microphone with sensitivity to frequencies and intensities of sound beyond those of the best human models. If we understood how it works, in the most intimate way, we might build a similar tiny, ultra-sensitive, biophysical-chemical system. With our increasing interest in miniaturization of sensitive systems, as we enter the space age, and our need to store vast amounts of information in small space, it is essential that we find smaller and smaller receiving and generating units. Biological units are these.

It is not alone in the nature of the receptor organs that the animal body is remarkable when compared with human information-handling systems. The animal nervous system is given the coded, pulsed, bioelectric messages created in the receptor by signals sent by another individual. In the central nervous system, these are decoded, sorted, and finally routed to appropriate reactive organs. Once again, this is done with structures of remarkably small size. The brain of a honey bee, for instance, weighs only a fraction of a gram, and occupies a space smaller than a pin-head, yet it has the ability to produce and interpret the complex signals that have already been discovered—and we are just beginning this study—and to profit by experience, through learning. Each neuron of this tiny system is like a data-storing or data-processing module of an electronic brain. Yet the electronic brain requires a large space to carry out processes that are less complex than those carried out by this remarkable piece of equipment that each honey bee carries around with it, and which disappears with it at its death.

Biological information-storing, coding, and sorting systems have another advantage over most human systems. They operate at full efficiency, although part of the system may be damaged. Large areas of the brain of a mammal, even man, can be destroyed, without noticeable effects on the ability of the individual to interpret signals and react appropriately. Large segments of signals may be taken away, without obvious reduction of their capacity to produce reactions. The reactive systems have many alternate pathways, and the messages are highly redundant. Contrast this with the situation in a human device, such as a radio or television set. Here the destruction of only one key element can nullify the working of thousands of others.

Once more, studies on how living sensitive systems handle information, how they code and decode signals, may show us how we can do the same efficiently, in a small space, yet have enough redundancy that the loss of one part does not destroy the action of the whole. Our so-called electronic brains are only brains in that they mimic a few of the more obvious characteristics of the output side of the human brain. To a biologist, the term, brain, is almost amusing when used for devices which have such weight and bulk for their load-level. No animal could afford to carry this much around. Furthermore, their lack of capacity to substitute one part for another automatically, within great limits, makes them a poor match for an animal brain. These two characteristics: small size and variable response-patterns, are among the most valuable that could be built into human sending and receiving systems. If we knew more about how the neurons of the nervous system worked, in a biophysical and biochemical sense, we might convert our crude, bulky imitations of animal nervous systems into actual homologues, with the desirable properties of living systems, but without the interfering vital activities.

WILDLIFE MANAGEMENT

In a more homely and direct way, knowledge of animal communication can be used by man in his recreational and agricultural pursuits. The use of decoys to attract wild animals is as old as the human race. Among the earliest records of hunting are accounts of the use of models and sounds of birds, such as ducks, to attract the birds for capture. Hunters still use decoys to induce reactions in wild animals so that they are easier to approach. With newer equipment for recording the sounds of animals, so that the meaningful parts of signals used in animal communication can be determined, we can now make these methods more precise.

As we have seen, many fish have odors that attract or repel other members of the species. When the chemical structures of the attractive materials are known, it is not too much to hope that these can be used in commercial fishing to attract fish so that they can be caught.

In French West Africa, natives used a sound-producing device, called the cotio-cotio, to attract fish. This has worked so well that some of these tribes do not use a fish-hook, a device that almost all primitive tribes have. The cotio-cotio is made of a piece of grooved bone, which is brushed by a piece of metal, thus producing a sound something like that made by a New Year's Eve rattle. The device is placed on the surface of the water and jiggled to make it rattle. The sound, transmitted through the water, attracts fish, and the native fisherman selects the most desirable and spears them. He has no need for hooks; his sound-lure works too efficiently.

As studies on the communication systems of fish progress, we can certainly look for increased use of the signals they produce in our attempts to improve sport fishing and to increase the catch of commercial fish so needed to feed the growing populations of man.

It is with birds and mammals that man has used his ability to mimic communication signals most extensively. Thus, by making models of birds, which generally act as visual signals to birds that food is present, he has attracted birds so that he can snare them. Furthermore, man has used whistles, or his own vocal apparatus, to bring the birds within range of his weapons. Primitive tribes, who have to make their living from the animals around them, often seem to understand the animals, and can thus influence their behavior. Among many primitive peoples, for instance, there are individuals whose sole job it is to go with hunters and call in birds or mammals with their voices. Although man may not be able to exploit wild bird populations extensively for food, he does, in civilized countries, hunt game as an important recreational activity.

Communication signals, furthermore, can be used in studies on populations of birds and animals by wildlife specialists. For instance, broadcasts of the distress call of Redwinged Blackbirds cause the birds to fly up from fields, so that they can easily be censused. It is difficult to count the birds if they stay down in the grasses and weeds. Undoubtedly, there will be many more applications of this sort.

To capture grouse for banding, so that their life histories and movements can be studied, the birds are attracted into a cage by using a mirror. The male grouse, seeing its image in the mirror, displays before the presumed invader and, losing all fear, goes right into the trap. Usually these birds are quite difficult to catch. Communication signals of birds, therefore, need not be used only to bring the birds to harm, but can also be used for the benefit of the birds.

With mammals, as with birds, there has been a long history of the use of communication materials for attraction so that man can capture them. Beaver trappers, during the early days of exploration of the western United States, used the sexually attractive extract of the glands of the beavers to make their traps attractive, or a least to eliminate avoidance reactions to them. Many hunters use fox calls, or calls of other predators, to attract the animals to their death. However, the signals can be used not only for these purposes, but also to aid in study and management of game mammals.

AGRICULTURE

In primitive farming, human beings often lived intimately with the animals, and understood the calls of the animals. This enabled them to keep track of the condition of their flocks and herds, and even, in some cases, to influence their behavior. We may expect, with continued research on communication in domestic animals, that the use of communication signals to influence their behavior will go even beyond the uses of the primitive agriculturists.

Discoveries on the social organization of chickens and turkeys, which use mainly visual and acoustical signals to maintain the flocks, have yet to be exploited to increase production on poultry farms. There is no doubt that this might be done. We shall give just one possible example. It has been the practice for some time, where chickens are kept for egg production, to have as few males present as possible, so that the eggs are not fertilized. This may not be as good a practice as it might seem, if a reaction recently discovered for parakeets is general. Female parakeets produce far fewer eggs, or sometimes none at all, if they do not hear the voices of males. A similar situation has been reported for grasshoppers, also—presence of males is necessary for full egg production by females. It could be, therefore, that removal of males from flocks of poultry may not be economically as smart as it might seem.

Experiments are now under way to see whether the use of sounds or visual signals during artificial insemination can increase the effectiveness of this technique in cattle and other animals. There is some evidence that the signals provided by bodily contact, or by the sounds and smells of the opposite sex, may improve the rate of fertilization. If these signals could be imitated artificially, avoiding the use of live decoy animals, a great deal of money could be saved.

Young animals that are reared separately from parents, for economic efficiency, often spend much time trying to communicate with their absent parents. The use of recorded auditory signals common between parents and young may bring about increased development of the young. Perhaps resemblances in sound may account for the fact that some animals do better and grow faster if music is played nearby. We have much to learn about the features in music that may resemble the communication signals of animals. Possibly, music that sounds soothing or happy to human beings may do so because it contains some underlying biological signals characteristic of all well-being calls. It might thus influence lower animals, as well as man.

Honey may no longer be an important item of man's diet, but much of it is still sold, and bees are still needed to pollinate some crop plants. Studies on communication among honey bees may enable us to use their complex systems for better management of the bees. Von Frisch has shown that honey bees can find their hives more easily when the hives are colored or shaped characteristically. This stops the bees' wandering accidentally into the wrong hive and being killed because they do not have the right colony odor. It would be interesting to know whether one could transmit information to bees in the hive through sounds or movements, and thus speed up their reaching desired crops or food sources. If one would want to have large numbers of bees working in an apple orchard, for instance, he might communicate to all the bees in the neighborhood, by electronically produced sounds and vibrations like those of the round or waggle dances, the location of the blossoms, and so bring the bees to the spot rapidly. Where weather factors interfere with pollination, as sometimes happens, this improvement of efficiency might be valuable.

The piping sounds of the queen honey bee, when received by the worker honey bees through their legs, cause the workers to stop short and remain quiescent. By vibrating the hive with sounds of certain frequencies, generally like the piping sounds of the queen, one can cause all the bees in the hive to remain immobile. These sounds, therefore, can be used by a beekeeper to still the bees while he opens a hive and takes care of them, without, as at present, using smoke. The smoke causes the bees to suck up honey in the hive and to spend considerable time afterward in restoring and ventilating. With sound, the activities of the hive are returned to normal as soon as the sound is turned off.

These are just a few possible uses of animal communication systems in agriculture and wildlife management. Only by much more research on the reactions of animals to signals that they produce, can we begin to formulate what might be done. The development of means for storing the communication signals artificially, with tape recorders or photographs, opens up new possibilities for the future.

PEST CONTROL

A pest may be defined as an animal that gets in man's way. Some animals, for instance many insects, get in man's way by feeding on his food or devouring his products. Other animals, such as some birds, get in man's way by occupying places he wants, or by producing discomfort or hazard by their presence. As man's civilization becomes more and more complex, and encroaches more and more on the natural haunts of animals, the animals get in man's way more and more; pest problems are increasing and will continue to do so.

Most persons, when they think of pests, think of those that are obvious competitors of man—rats and mice, which eat stored food; crows and blackbirds, which eat growing grains; and insects, such as locusts and mosquitoes, which attack crops or even man himself. In these cases, man realizes that their competition may endanger his life, and he is willing to kill them off, if need be. Thus he does not hesitate to use rat poisons, guns, or insecticides in his attempts to save his crops, his foods, or his own person.

Many other so-called pests, however, are only pests at certain times, or under certain circumstances. This is true in the case of most birds. For instance, Starlings might seem to be unmitigated pests in the United States, because they gather in large, defiling flocks on public buildings or trees, and they raid feed-lots or cornfields. But this is by no means true. The food habits of Starlings, particularly during the breeding season, benefit man. At this time, they eat large masses of harmful insects; actually, they eat quantities of harmful insects throughout the year. Any attempt to poison or kill large groups of Starlings would almost certainly endanger other species of birds with similar food preferences, and this of course we do not wish to do. In these cases, therefore, we should be able to control the habits of the birds without harming them.

With most pest birds, the situation is even more anomalous than it is with Starlings. For instance, woodpeckers are generally considered to be valuable and worth saving. Yet, in some places, woodpeckers are nuisances because of their drumming, or are pests, because they peck holes in power-poles. An electric power company has no wish to destroy woodpeckers, particularly where they are relatively rare, but the company would like to save its power-poles.

Birds have learned to adapt themselves to civilization in peculiar ways. Small birds in Europe have learned to peck the tops out of milk bottles to get milk. Others have found rubber blades from windshield wipers on automobiles useful in building their nests. After all, man has taken the usual sources of these necessities from the birds, and they will not simply die. In these cases, it would be most valuable if we could control the movements and behavior of the animals without hurting them.

In practical terms, pest control has been attempted with two sets of communication signals which seem to have the most promise—chemical signals, and acoustical signals. Very seldom can visual signals be used efficiently, although we must not forget the many types of scarecrows used for chasing birds.

Chemical controls

Among chemical signals, the odor produced by the female gypsy moth has been used to attract the males to their death. The odor of female codling moths has been used to attract males for censusing the population, to get data on when to apply insecticides. The attraction of male cockroaches to the odor of females indicates that odor could be used to attract the males to their death. Unfortunately, in most of these cases, only one sex is attracted, and, if it is the male, the control is apt to be ineffective in reducing the population, for some males remain to fertilize females, and female egg production is usually very high.

Chemo-sterilants, that is, chemicals which cause sterility in insects, are attracting considerable interest as agents for insect control. Obviously if these materials are to be used on crop plants they must not harm man or domesticated animals. A natural chemo-sterilant is the queen substance of honey bees. The pure queen substance, while suppressing the development of sexual maturity in female honey bees, has been reported by its discoverers to have no untoward effects on the mammals on which it was tried. Possibly, since it is a naturally produced substance, animals, in general, would be able to take care of it within their bodies. We are not usually interested in inhibiting reproduction in useful insects, such as honey bees, and we are willing to let them use queen substance for their own social integration. However, this might be only one of a number of similar inhibitory substances which could reduce or completely inhibit reproduction in other insects.

Many mammals are known to have, or are suspected of having, alarm odors. Anyone who has worked with laboratory rats or mice, and who has used his sense of smell appreciatively, knows that these animals release acrid odors when they are maltreated. There is no question that other members of a rat or mouse colony respond to these alarm odors. It is a matter of common knowledge that a trap in which a rat or a mouse has been caught may not catch another rat or mouse for some time thereafter, probably because of an alarm odor. We still know little about the odors produced by mammals, and much research should be done on these. It might be that some odors of mammals other than man would be odorless to man, and so could be used around human habitation to keep these animals away.

It has been reported that deer have an alarm scent which is given off by a special set of glands. Deer are pests to farmers in many parts of the world, because they invade orchards and nibble at the tender ends of tree branches. No one wants indiscriminate slaughter of these valuable game animals, yet the farmer cannot see his orchards destroyed. If he could apply an alarm scent from the deer to repel them from the orchards, it would be a boon both to man and deer.

Acoustical control

The development of the tape recorder and modern public address equipment has made possible the storing of acoustical communication signals of animals and later broadcast of them to the animals in the field. This has led to the first successes in practical control of pests by communication signals and has tremendous potentialities. The work is just beginning, and is still mostly a matter of possibilities.

We have noted the many insects that produce sounds for communication purposes and the wide variety of behavior patterns brought about by these signals. It certainly would seem that the time has come for an all-out effort to see whether these can be used, even with the little bit we already know, to influence the behavior of insects for our benefit.

FIGURE 24. *Attraction of birds by broadcasting recorded assembly call, like that given by crows on sighting an owl or cat.*

Sounds of very high intensity can kill insects, but the costs are too high to be practical. It seems more practical to use sounds either to attract insects to their death, or to chase them away from man or his crops. Attraction of insects by sounds, however, does not offer much promise, for attractive sounds are usually sexual sounds, and so would have to compete with the natural sounds used in communication between the sexes.

This is well illustrated by the situation in mosquitoes. It has been suggested that the recorded sound of the wings of female, disease-carrying mosquitoes be used to attract the males. However, only some of the males are attracted, and the others are left to fertilize the females. If a male mosquito were nearer to a female than to an artificial sound source, the male would be attracted to the female and could fertilize her. To create a situation in which the artificial sound would be louder than any female in the region would require myriads of speakers. One cannot merely increase the intensity coming from one speaker, for, above a certain level of intensity, the sound is no longer attractive, but repellent. Since female mosquitoes are capable of producing tremendous numbers of eggs after one fertilization, and the major controlling factor in the population of mosquitoes is the number of larvae dying because there is not enough food for them, just a few fertilized female mosquitoes can maintain the population at its normal level. Thus, attracting the males to their death, even if efficient at the 99% level, would have no effect whatsoever on the mosquito population. Since the males do not bite, and therefore do not transmit disease, the immediate reduction in their numbers would be useless.

Conceivably, if one could attract and destroy both sexes, or attract the sex that does the damage, some alleviation of the trouble would be obtained. It is possible, for instance, to attract large numbers of both males and females of some insects to lights, and so to take them away from prospective food. Again, however, the natural attractant is competing with the artificial attractant, and, because generally the artificial stimulus cannot be of unlimited intensity, the attraction is only partial.

On the other hand, repellent sounds, such as alarm signals, offer more hope for practical control. Thus, if flies or mosquitoes were found to use characteristic, specific, warning sounds, these could be recorded and broadcast to the insects to keep them away from food, or from man himself. So far, there has been too little work to predict how effective this might be.

As we have seen, many insects have acoustical signals, and it is only a matter of time until critical tests will be made to see whether these can affect their behavior for man's benefit. Among the economically important forms that have well-developed acoustical communication systems are the following: grasshoppers, with very elaborate stridulatory languages; mosquitoes, in which the females attract the males; other flies, which also use wing-sounds for sexual attraction; and numerous beetles, which use stridulating sounds, probably for sexual communication. In all these cases, however, much more information is needed before we shall be able to predict what might be done.

The first practical pest control using acoustical signals has been accomplished with birds. When a Starling is held by the legs, it emits a raucous distress call. This has been recorded, and when played back to Starling flocks, causes them to fly away. Obviously, this is not enough; one must keep the birds away over an extended period of time. How long depends upon what the birds are doing: if they are feeding on crops, one may need to keep them away for days or weeks; if they are roosting at night, one may only need to chase them while they are trying to fly into the place where they mean to spend the night.

To control roosting Starlings, the recorded distress call is broadcast in the evening, as the birds try to enter their place of rest. This causes them to fly away, thus disturbing their highly stylized roosting pattern. Usually, if the disturbance is kept up for four or five successive nights, as the birds try to come into the roost, they desert the roost and find some other place where there is less disturbance. Actually, one can use various noises and occasionally achieve the same result. Usually, however, the birds learn that the noises do not mean imminent danger, and they have no other significance to them. The reaction to the distress call is a built-in reaction that enables the species to remain alive, and so the birds generally continue to respond.

Interestingly enough, Starlings may learn that the broadcast distress call is phony, if one plays it for too long at one time. The birds probably detect that the call is being repeated many more times than normal, or are able, with repetition, to note slight differences between the recording and the real thing. In controlling roosting Starlings, therefore, one uses the distress call for as short intervals as possible, turning the call off as soon as the birds have deserted the spot.

Experiments are now in progress, in many parts of the world, on the use of the recorded distress call of Starlings to keep these birds away from buildings, airport runways, or crops. Naturally, if this method is to be used economically, one should be able ultimately to have the calls broadcast in the correct temporal patterns fully automatically. This will require the study of the movements of the birds to determine the criteria for repetition of the sounds.

Studies with other birds also show considerable promise. Thus, Crows can be either attracted by an assembly call, or repelled by an alarm call, allowing control of their movements. In France, extensive experiments have been made on the control of crows feeding on crops, and at their roosts or nests. The French workers have shown that, where there is need for population reduction, the calls can be used for this. Distress or alarm calls, broadcast in the middle of the night, drive the birds from their nests, leaving their eggs and young. In the darkness, the adults do not find their way back, and the eggs and young perish in the coolness of the night.

Herring Gulls and Blackbirds also can have their behavior influenced by sound, and studies are now under way to determine the best way to use recorded calls for the control of these birds around airports, or where they feed on corn or other crops. The Herring Gull, for instance, has a fairly large vocabulary, including an attractive call, the food-finding call; a repellent call, the alarm call; and a distress call. It would certainly seem that combinations of these could be used to solve the problem created by the activities of these large birds near airports, where they collide with planes. Experiments in England and in Holland indicate that this may soon be a practical reality.

The practical use of recorded bird calls to control the movements of birds has been a recent development, and relatively few individuals have worked on it. With increasing interest and greater capabilities of equipment, the management of animal populations by acoustical communication systems should become a reality. This should benefit both man and birds.

Modern work on control of mammals with sounds is scant. As with birds, people have used loud noises, such as those made by exploders or guns, to chase mammals from places where they are not wanted. These methods have proved to be rather expensive, and the animals rapidly learn that there is no danger. It is almost certain that shooting at mammals and birds is repellent solely because shots are loud noises. Thus, recording the shots is a much more efficient and economical way to get to the pests. However, the rapid adaptation of mammals to mere noises makes them relatively useless. As with birds, communication signals would seem to be the answer, but very little has been tried.

One of the objections to the use of recorded distress calls to chase birds has been that the calls are audible to man, and so, when broadcast at fairly high intensities, may be almost as repellent to him as to the birds. However, many mammals that are pests do their damage away from settled areas. Thus, deer in orchards are usually far enough from man that loud sounds used to chase them would not reach human ears.

The acoustical communication signals of rats and mice, the most important mammalian pests, are often ultrasonic for man, that is, the animals use frequencies of sound too high for the human ear to hear. This would allow us to intrude into their communication systems and to broadcast even at high intensities without man's being even able to sense it. Interestingly enough, there are many folk tales about rats and mice being influenced by sound; the most familiar, of course, is the story of the Pied Piper of Hamelin. So far, studies on communication systems in the ultrasonic range have been delayed, because commercial sound equipment is generally not adapted to handle these frequencies. Investigations into the communication channels of rodents, which are ultrasonic and thus inaudible to man, should yield some interesting practical results.

Again and again, in this discussion we have come back to the point that, for possible practical utility, it will be necessary to have information which we do not now have. In many cases, we do not even know what this information should be. Thus, at this stage the most practical approach to the use of animal communication signals for man's benefit might seem the least practical: that is, to continue to get fundamental information, regardless of whether it seems to have any immediate practicality or not. Only with a broad base of fundamental knowledge can we begin to predict with any accuracy the possibilities for practical uses.

As attempts are made to turn our rudimentary knowledge of animal communication to practical purposes, further pioneering research will be engendered. Attempts to use communication signals in pest control or in agriculture, may, because of the lack of fundamental knowledge, catalyze an interest in pure research that will give us more fundamental knowledge and enable us to understand better the basic nature of animal communication. Throughout science and engineering, persons interested in practical and fundamental problems thus interchange information and inspiration to the benefit of all.

11

The Future

ONE would be rash indeed, in science, to try to predict what the future will bring. Obviously, if one knew what discoveries were soon to be made, he could make them now, and then there would be others to be made. Many cases could be cited of persons who attempted to make predictions only to find that, before the ink was dry on their words, some new discovery invalidated all.

We might just insert a personal item along this line. In May, 1953, in a talk at the Pasteur Institute in Paris, we were asked to comment on the possibility of using sounds for bird control. At that time, all that we and others could think of was the use of loud noises, or noises of some type at least, and it was obvious that birds easily adapted to these. We reported, therefore, that we could see nothing that seemed the least bit hopeful for alleviating the bird pest problem.

At the very moment when this statement was made, we had, at the Pennsylvania State University, a recording of the distress call of the Starling, which we had made in March. This had been recorded, because a young man, Joseph Jumber, who was working with us, had observed that, when a Starling emitted this call, others in the neighborhood left precipitously. It seemed possible, therefore, that this might have some use for chasing the birds.

We thought so little of it, however, that nothing was done to try this out until August of the same year.

We have already reported the success with which our efforts were crowned. It is not enough, however, to use the distress call indiscriminately. It is absolutely necessary to use it with regard to the behavior of the birds, to insert ourselves, so to speak, into their communication systems. With this knowledge, practical control for Starlings has been devised, under appropriate circumstances.

What was needed here was an awakening to the fact that mere noises, with no biological significance to animals, cause reactions only by startling, whereas communication signals touch the animals at the basic roots of their social structure. Animals can learn that noises are harmless, but they dare not, at the peril of their lives, ignore communication signals, such as alarm or distress calls. A new concept was needed, and when this concept was found, a whole new world of research opened up. Had we had the concept in May of 1953, obviously we would have had the discovery already made.

If, however, we can make no predictions that are more than guesses, we can at least make some guesses as to what future research on animal communication will divulge. We can, for instance, note developments which have already started and are almost certain to continue.

NEW USES FOR EXISTING EQUIPMENT

As biologists become more familiar with the newer equipment available to them, they should be able to use it more efficiently. The tape recorder, as a practical unit, is now only about fifteen years old. Before 1945, the recording of sounds was a troublesome and expensive process. Now tape recorders are so inexpensive and small that they can be used by almost anyone who is interested in research on animal sounds.

We should see, in the next few years, much more imaginative use of the tape recorder. One could think, for instance, of the possibilities of playing tapes at different speeds, and of using recordings variously dissected and inverted, so that one could test the meaning of individual portions of bird or insect songs. Commercial recording engineers routinely remove a note in the middle of a symphony and insert another one that is more to the liking of the performer. There are few biologists who have tried to develop this technique with animal sounds, but we should see it, as we ask much more critically what are the important features of the sound signals.

Tape recorders are generally so engineered that they work best with human voice and music. Many insect sounds, however, are quite different in their physical characteristics from the sounds of human voice or music, and handling these—to have them hi-fi for insects—may require re-engineering of tape recorders for this particular use. Also acoustical signals of many insects and some mammals are ultrasonic, and generally our equipment is not designed for this range, for it is a range which man does not hear. It is to be hoped that possible practical utility of sounds in this range will induce manufacturers to produce tape recorders and associated equipment, such as microphones, which will work efficiently in the ultrasonic range, and so handle the sounds of insects and other animals.

In the study of visual communication patterns, it is necessary not only to record the patterns of the animals, which can be done quite easily with a color camera, but also to record the use of these patterns in the complicated movements of the animals. These can, of course, be photographed as moving pictures; there are already some collections of sound movies showing communication behavior in animals. However, film is expensive, and generally the person must be fairly sure, before he runs long stretches of film, that he is going to get what he wants. This may be contrasted with the situation in the case of the tape recorder. The recorder can run almost indefinitely, and the tapes can later be edited to pick out the parts that are important. Only these sections need be saved; the rest of the tape can be reused, again and again.

The development of the T-V tape recorder offers students of animal behavior a new, extremely valuable tool. This device records on magnetic tape, as does the sound recorder, and the tape, therefore, can be used over and over. At present, the limiting factor is the high cost of the recorders. It is an interesting commentary on comparative values that expenditures of large sums of money to buy T-V recorders can be defended for advertising purposes, but not for equipping laboratories to study animal behavior. It is to be hoped that, in the near future, the price of T-V recorders will be within the resources of laboratories interested in monitoring animal behavior over long periods of time.

DEVELOPMENT OF NEW TECHNIQUES

Many aspects of the study of animal communication will require the development of new techniques. Some will be merely improvements on older techniques, but improvements enough to be significant. Others, however, will have to be definitely new.

Almost every new technical discovery in chemistry and physics could conceivably be useful to students of animal behavior. For example, the development of chromatographic methods, particularly gas chromatography, has allowed biochemists to separate and to characterize compounds forming the odor and taste signals of insects. Newer refinements and developments of these methods should enable us to do this even more efficiently.

As in the case of tape recorders, these methods should enable us to distinguish the effective parts of signals from those that are present more or less accidentally, or to determine the nature of redundancy, the different materials carrying the same information. Attempts to discover the essential fundamental units in the informational systems should be of extreme interest not only to biologists, but also to biophysicists and biochemists. These scientists have long used, in their own disciplines, the analytical method—breaking down a complex situation into the simplest units, and then studying them one at a time.

For the study of communication by taste and smell, the development of some system of classification of tastes and smells, and the ability to store and collect them, would be of tremendous value. We need, for instance, the development of a smell recorder to collect, store, and play back the symphonies of odors which animals produce, as visual or acoustical signals can be manipulated. It is probable that the development of such a device will involve not the transformation of the odor into an electrical or mechanical signal, as in optical or acoustical recording, but the transmutation of the odor into some other chemical arrangement which can be stored and later transposed back into the odor signal. Man's relative indifference to odors, which for most animals constitute probably the major communication channel, is shown by the fact that progress in this field is far behind that in the study of optical and auditory systems.

As we have already mentioned, the new field of bionics should become of more and more interest to biologists, physicists, and chemists. This should mean an interest in all phases of communication, but particularly in the determination of the intimate nature of the signals involved. As physicists and chemists become interested in biological phenomena, they should realize the inadequacies of some of the devices which work well in their particular fields when these are used to study such complex systems as those of the living world. This should bring about the development of new devices and techniques. Whenever any significant breakthrough occurs in the understanding of biological systems, through the understanding of their physical and chemical bases, interest should grow by leaps and bounds. Development of a practical application for discoveries in animal communication should attract many people, who otherwise would not be interested, to study communication intensively.

As man becomes more and more interested in storage, coding, and retrieval of information, he should become more and more interested in how this is done in the living world, for this, in essence, is what an animal brain does. The rather crude models of the living brain, called electronic brains, have already given us some indications of the possible systems available for animals. As studies progress on the actual systems used by animals, it is not too much to hope that new principles for design of electronic brains and computers will be discovered. There should be mutual feed-back between technologists and biologists. With this, interest will center again on the effective features of animal signals and the limits of informational transmission and storage in the channels that are open to animals. This should have a desirable result: forcing man from his preoccupation with only a few communication channels—those which his senses use best—to broader prospects.

BIOLOGICAL ENGINEERING

If communication signals can be used to solve some of the practical problems facing man, the biological engineer will take his place alongside the electrical engineer, civil engineer, and so on. The latter take the fundamental discoveries of one of the sciences, such as physics or chemistry, and apply them to the solution of practical problems. They may, at the same time, need to discover new information to aid them in making the applications, but primarily they hope to find the necessary information already at hand.

In a sense, medical men and agricultural specialists are biological engineers, utilizing knowledge derived from biochemistry, physiology, and biology to solve problems that arise in particular fields of animal life. However, these are restricted fields, and, if further similarities between animal communication and human communication are discovered, or particularly if the sensitive systems used by animals give information that can be used to produce sensitive systems for human use, we should see the development of biological engineers who will take biological information and apply it for man's benefit.

In the broader view, this should apply not only to animal communication and behavior, but to all biology, for there are important biological principles in genetics, ecology, and many other fields, whose potential applications to human welfare are immeasurable. Biological engineering should then come into its own as a recognized profession, along with other types of engineering where work is founded on fundamental chemistry and physics.

For a biologist working in the field of animal communication the future is one that should hold numerous and often startling surprises. Research here is often of the most pioneering type. Some research can be fairly well plotted out, and one can be fairly sure, if he carries out the work the way he has planned it, that, within certain limits, he is going to get certain results. In pioneering research, on the other hand, it is usually impossible to predict what is going to happen. Research here is much like exploration of a totally unknown area of the universe. One may guess what he is going to find, and plan in the hope that he will be able to find it. But only as he begins to penetrate into the unknown does he really discover what is there, and this may be quite different from what he thought. Research in animal communication, therefore, should offer intellectual adventure of the highest order.

NEEDS IN THE SPACE AGE AHEAD

The world, our earth, is an almost infinitely vast universe for most of the animals that live on it. Systems of communication have been developed enabling animals to identify their species and to carry out the processes needed to keep the species in existence.

As man now enters the space age and begins to invade a universe that is almost infinitely vast, even at his size scale, he will find himself in much the same position as these tiny mites of life which he sees every day on earth. He will be faced with vast distances in which he must be able to find other members of his species. Thus he will be faced with the problem of communicating with them.

He may even be faced with the prospect of finding, some where in the universe, animals that have developed to the same —or an even higher—level of information acceptance and storage capacity than he. These may, however, be using channels other than those used by man. It may be necessary to understand these other channels, and to be able to carry out rudimentary communication in them.

Thus, man has reached the stage where knowledge of animal communication may not only help him to solve some of his everyday problems, but may also be necessary to aid him as he probes into the infinity of space. Here he may need to have a variety of efficient, distance-spanning, communication systems. He may need, too, the built-in safety features of animal communication systems, so that no mistake in identity or purpose can be made.

It now behooves man to learn more about the ways in which his smaller relatives communicate with each other. Not only, as a civilized animal, interested in knowledge for its own sake, should this be meaningful to him. What he can learn about animal communication may determine his ability to survive and prosper in a universe that is rapidly becoming for him—as for his tiny relatives—almost infinitely vast.

Bibliography

THE books and articles listed here form a selected set of references, mostly of a general nature, often giving further references to original literature. The works cited represent only about 5–10% of those used in the preparation of this book. Undoubtedly, another and different list of equal length could be prepared, which would include many other works that would be considered important. In short, this is not a complete or even definitively selected bibliography.

The following words are used, in the descriptive statements following the references, with special meanings: *technical*—written for the research specialist, requiring knowledge of a specialized vocabulary; *nontechnical*—written for the intelligent and interested layman; *popular*—written for the general public, carefully avoiding technical terms.

If the descriptive statement is followed by the word Biblio., the book or article has a large bibliography. The numbers in parentheses after each statement refer to the chapters of this book for which the reference is pertinent; if the reference deals with the subject of animal communication in general the word General is used.

Alexander, Anne J. 1959 Courtship and mating in the Buthid scorpions. Proc. Zool. Soc. London, 133: 145–169. A technical report of behavioral observations and descriptions of mating organs. (7)

Alexander, Richard D. 1962 Evolutionary change in cricket acoustical communication. Evolution, 16: 443–467. A technical review and research report on the evolution of mating sounds in crickets. (4,6,7,9)

Allee, W. C. 1951 Cooperation among animals, with human implications. Rev. Ed., First Ed. 1938. 233 pp. Henry Schuman, New York. A nontechnical report on research on the social interactions in feeding and flocking. Biblio. (3,4,5)

Armstrong, Edward A. 1947 Bird display and behavior. 431 pp. Oxford Univ. Press, New York. A general nontechnical review of all aspects of social behavior, particularly sexual behavior, of birds. Biblio. (4,6,7,8,9)

Armstrong, Edward A. 1963 A study of bird song. xv + 335 pp. Oxford Univ. Press, London. A technical review of all aspects of bird-song, with a short section on communication in the Animal Kingdom. Biblio. (General)

Beament, J. W. S. (Ed.) 1962 Biological receptor mechanisms. Symposia of the Society for Experimental Biology, No. 16. vii + 372 pp. Technical reports on research in various aspects of sensory physiology by a number of workers. (2,3)

Blair, Frank 1958 Mating call in the speciation of anuran amphibians. American Nat., 92: 27–51. A technical report of research on calls of frogs and toads as mating signals. (3,6,9)

Bonner, John Tyler 1959 Differentiation in social Amoebae. Sci. American, 201(6): 152–158; 160; 162. A nontechnical report of research on communication in slime-molds. (4,6)

Bonner, John Tyler 1963 How slime molds communicate. Sci. American, 209(2): 84–86; 89–91; 93. A nontechnical report of research. (4,6)

Bristowe, W. S. 1958 The world of spiders. xiii + 304 pp. Collins, London. A nontechnical presentation of knowledge of all phases of spider life, including much on behavior. (4,6,7)

Buck, John B. 1948 The anatomy and physiology of the light organ in fireflies. Ann. New York Acad. Sci., 49: 397–482. A technical review of this subject, with a report of original researches. Biblio. (2,3,6)

Burton, Maurice 1953 Animal courtship. 267 pp. Frederick A. Praeger, New York. A popular general survey of this subject. (6,7)

Busnel, René-Guy (Ed.) 1955 Colloque sur l'acoustique des Orthoptères. 448 pp. Fascicule hors série des Annales des Épiphyties, Exemplaire No. 6. Institut National de la Recherche Agronomique. Paris. Technical reports by a number of workers on acoustical communication in crickets and grasshoppers. Biblio. (2,3,4,6,7,9)

Busnel, René-Guy 1963 Acoustic behavior of animals. xx + 933 pp. Elsevier Publ. Co., Amsterdam. A series of technical papers by 23 authors reviewing all phases of bio-acoustics. Biblio. (General)

Butler, Colin G. 1949 The honeybee—an introduction to her sense-physiology and behaviour. vi + 139 pp., Clarendon Press, Oxford, England. Review of research on this social insect, especially behavior and communication. Biblio. (4,5,7,8)

Butler, Colin G. 1954 The world of the honeybee. xvi + 226 pp. Collins, London. A nontechnical general review of knowledge of the life of this social insect. (4,5,7,8)

Butler, C. G., R. K. Callow, and Norah C. Johnston 1961 The isolation and synthesis of queen substance, 9-oxodec-*trans*-2-enoic acid, a honeybee pheromone. Proc. Roy. Soc. B, 155: 417–432. A technical report of research on queen substance of bees. (3,8)

Carthy, J. D. 1958 An introduction to the behaviour of invertebrates. viii + 380 pp. George Allen & Unwin Ltd., London. A technical survey of the senses, and briefly, behavior of invertebrates. Biblio. (2,3)

Cherry, E. Colin 1952 The communication of information. American Sci., 40: 640–664. A theoretical, technical discussion of communication theory and human and animal language. (1,9,11)

Cloudsley-Thompson, J. L. 1958 Spiders, scorpions, centipedes, and mites. xiv + 228 pp. Pergamon Press, London. A technical review of the ecology and natural history of these animals. (4,6,7,8)

Crane, Jocelyn 1949 Comparative biology of Salticid spiders at

Rancho Grande, Venezuela. Part IV. An analysis of display. Zoologica, 34: 159–214. A technical report of observations on sexual communication in jumping spiders. (4,6,7)

Crane, Jocelyn 1957 Basic patterns of display in fiddler crabs (Ocypodidae, Genus *Uca*). Zoologica, 42: 69–82. A technical report of observations on sexual behavior of these animals. (3,4,6,7,9)

Dilger, William C. 1962 The behavior of lovebirds. Sci. American, 206(1): 88–98. A nontechnical report of observations on sexual activities of lovebirds, including visual and auditory communication. (3,4,6,7,9)

Ehrlich, Paul R. and Richard W. Holm 1963 The process of evolution. xvi + 347 pp. McGraw-Hill Book Co., Inc. New York. A technical general presentation of modern ideas on evolutionary mechanisms; no special discussion of behavior, however. (9)

Eibl-Eibesfeldt, Irenäus 1961 The fighting behavior of animals. Sci. American, 205(6): 112–116; 119–120; 122. A nontechnical review of fighting and ritualized fighting by vertebrates of all classes. (4,5)

Faber, Albrecht 1953 Laut- und Gebärdensprache bei Insekten Orthoptera (Geradflügler). Teil I. Vergleichende Darstellung von Ausdruckformen als Zeitgestalten und ihren Funktionen. Stuttgart, 198 pp. An elaborate technical classification of the sounds produced by European grasshoppers—about 400 types described. (5,6,7)

Frings, Hubert, Mable Frings, Beverley Cox, and Lorraine Peissner 1955 Auditory and visual mechanisms in food-finding behavior of the herring gull. Wilson Bull., 67: 155–170. A technical report of research. (2,3,5)

Frings, Hubert and Mable Frings 1958 Uses of sounds by insects. Ann. Rev. Ent., 3: 87–106. A brief technical review of research on insect hearing, sound production, and acoustical communication, 1950–1957. Biblio. (3,4,5,6,7,10)

Frings, Hubert and Mable Frings 1959 The language of crows. Sci. American, 201(5): 119–120; 122–124; 126; 128; 130–131. A nontechnical report of research on acoustical communication in crows, showing dialects. (5,9)

Frings, Hubert and Mable Frings 1962 Pest control with sound. Part I. Possibilities with invertebrates. Sound, 1: 13–20. A non-technical review of the use of sounds, particularly communication signals, for control of invertebrate pests. (10)

Frings, Hubert and Mable Frings 1963 Pest control with sound. Part II. The problem with vertebrates. Sound, 2: 39–45. Same as above for vertebrates.

Frisch, Karl von 1950 Bees—their vision, chemical senses, and language. xiii + 119 pp. Cornell Univ. Press, Ithaca, New York. A short nontechnical report of research. (2,4,5)

Frisch, Karl von 1962 Dialects in the language of the bees. Sci. American, 207(2): 78–80; 83–84; 86–87. A nontechnical report of research on various races of bees. (3,5,9)

Gertsch, Willis J. 1949 American spiders. xiii + 285 pp. D. Van Nostrand Co., Inc., New York. A nontechnical review of the general biology of spiders, with special coverage of American species. (4,6,7)

Hafez, E. S. E. (Ed.) 1962 The behavior of domestic animals. xiv + 619 pp. Baillière, Tindall & Cox, London. Technical reports of research on this subject, including communication, by a number of workers. Biblio. (3,5,7,8,10)

Haskell, P. T. 1961 Insect sounds. viii + 189 pp. Quadrangle Books, Chicago. A technical review of insect sound production, reception, and acoustical behavior, in all aspects. (General)

Hinde, R. A. 1959 Behaviour and speciation in birds and lower vertebrates. Biol. Rev., 34: 85–128. A technical review of research on mate selection, food selection, and habitat selection in these animals. Biblio. (6,7,9)

Kaston, Benjamin Julian 1936 The senses involved in the courtship of some vagabond spiders. Ent. Americana, New Ser., 16: 97–167. A technical report of research on these animals. (3,6,7)

Kendeigh, S. Charles 1952 Parental care and its evolution in birds. Ill. Biol. Mono. Vol. 22, Nos. 1–3. x + 358 pp. A technical review and report of his experiments on territorial and brood-care. Biblio. (7,8)

Klopfer, Peter H. 1962 Behavioral aspects of ecology. 166 pp. Prentice-Hall, Inc., Englewood Cliffs, New Jersey. A nontechnical survey of prey-predator relationships, species diversity and distinctness, and community organization. (4,5,6,9)

Lanyon, W. E. and W. N. Tavolga (Eds.) 1960 Animal sounds and communication. xiii + 443 pp. Publ. No. 7, American Institute of Biological Sciences, Washington, D. C. Technical reports and reviews of studies on acoustical communication by birds, insects, fish, frogs, and toads, with some general, theoretical discussions. (General).

Limbaugh, Conrad 1961 Cleaning symbiosis. Sci. American, 205(2): 42–49. A nontechnical report on methods used for identification by animals that clean others. (5)

Lindauer, M. and W. E. Kerr 1960 Communication between the workers of stingless bees. Bee World, 41: 29–41; 65–71. A semitechnical report of research and observations on guidance to food in relatives of the honey bee. (4,5,9)

Lindauer, Martin 1961 Communication among social bees. ix + 143 pp. Harvard Univ. Press, Cambridge, Mass. A nontechnical presentation of observations and experiments. (4,5,8,9)

Lindauer, Martin 1962 Ethology. Ann. Rev. Psychol., 13: 35–70. A brief technical review of research, mostly during 1955–1961, on social life and communication among animals. Biblio. (4,5,6,7)

Loewenstein, Werner R. 1960 Biological transducers. Sci. American, 203(2): 98–106; 108. A nontechnical report on experiments on touch receptors of vertebrates to elucidate the method for production of nerve impulses. (2,3)

Marler, Peter 1959 Developments in the study of animal communication, pp. 150–206, in: Darwin's biological work—some aspects reconsidered, Ed. by P. R. Bell *et al.* Cambridge Univ. Press, Cambridge, England. A semitechnical review of recent research on aspects of communication originally studied by Darwin. Biblio. (General)

Marler, Peter 1961 The logical analysis of animal communication. Jour. Theoret. Biol., 1: 295–317. A theoretical, technical discussion of animal communication in comparison with human language. (1,9,11)

Martof, Bernard S. 1961 Vocalization as an isolating mechanism in frogs. American Midl. Nat., 65: 118–126. A technical report of laboratory research on responses of female frogs to recorded calls of males. (3,6,9)

Matthews, L. Harrison and Maxwell Knight 1963 The senses of animals. 240 pp. Philosophical Library, New York. A nontechnical review of field and laboratory studies. (2,3)

Mayr, Ernst 1963 Animal species and evolution. xiv + 797 pp. Harvard Univ. Press, Cambridge, Mass. A technical general coverage of animal evolution in most aspects, including sections on behavior. (9)

McElroy, William D. and Howard H. Seliger 1962 Biological luminescence. Sci. American, 207(6): 76–87; 89. A nontechnical review of studies on light production by plants and animals. (2,3,6)

Milne, Lorus J. and Margery J. Milne 1954 The mating instinct. 243 pp. Little, Brown and Co., Boston. A popular survey of mating habits of animals. (6,7,8)

Milne, Lorus and Margery Milne 1962 The senses of animals and men. x + 305 pp. Atheneum, New York. A popular report of knowledge of animal senses and their uses. (General)

Moynihan, M. 1962 Hostile and sexual behavior patterns of South American and Pacific Laridae. Behaviour, Suppl. VIII. xiii + 365 pp. A technical report of observations on calls, postures, et cetera of sea gulls of these regions. (4,5,7,8)

Nicol, J. A. Colin 1960 The biology of marine animals. xi + 707 pp. Sir Isaac Pitman & Sons Ltd., London. A technical review of this broad field, including chapters on sense-organs and behavior. Biblio. (2)

Nicol, J. A. Colin 1962 Animal luminescence, pp. 217–273 in: Advances in comparative physiology and biochemistry, Ed. by O. Lowenstein. Academic Press, London and New York. A technical review of studies on the mechanisms and uses of luminescence of animals. Biblio. (3,6,7)

Pierce, George W. 1948 The songs of insects—with related material on the production, propagation, detection, and measurement of sonic and supersonic vibrations. 329 pp. Harvard Univ. Press, Cambridge, Mass. A technical book detailing analyses of the sounds produced by insects. (3,6)

Portmann, Adolf 1961 Animals as social beings. 247 pp. A

popular presentation of the forms of social life, organs of communication, courtship, et cetera of animals. (2,4,5,6,7,8)

Ribbands, C. R. 1953 The behaviour and social life of honeybees. 352 pp. Bee Res. Assoc. Ltd., London. A technical general review of knowledge in this field. Biblio. (5,6,7,8)

Roe, Anne and George Gaylord Simpson (Eds.) 1958 Behavior and evolution. viii + 557 pp. Yale Univ. Press, New Haven. Technical reports by a number of workers on the evolutionary importance of animal behavior in general, including a number on communication. (General)

Rosenblith, Walter A., Ed., 1961 Sensory communication. xvi + 844 pp. John Wiley & Sons, New York. A series of technical papers presented at a symposium on sensory physiology. (2,3,11)

Schwalb, Hans Helmut 1961 Beiträge zur Biologie der einheimischen Lampyriden *Lampyris noctiluca* Geoffr. und *Phausis splendidula* Lec. und experimentelle Analyse ihres Beutefang-und Sexualverhaltens. Zool. Jahrb., Abt. Syst., 88: 399–550. A technical report of observations and experiments on behavior of fireflies. Biblio. (3,6,7)

Scott, John Paul 1958 Animal behavior. xi + 281 pp. University of Chicago Press, Chicago. A semi-technical textbook on the general subject, with a chapter on communication. (General)

Shaw, Evelyn 1962 The schooling of fishes. Sci. American, 206(6): 128–134; 137–138. A nontechnical report of research on the senses and behavior patterns involved. (3,4,5)

Spieth, Herman T. 1952 Mating behavior within the genus *Drosophila* (Diptera). Bull. American Mus. Nat. Hist., 99: 401–474. A report on original research on courtship and mating in the fruit fly. (3,6,7,9)

Tavolga, William N. 1956 Visual, chemical and sound stimuli as cues in the sex discriminatory behavior of the Gobiid fish *Bathygobius soporator*. Zoologica, 41: 49–64. A technical report of original observations, and a review of related studies on other fish. (2,6,7)

Tembrock, Günter 1959 Tierstimmen—Eine Einführung in die Bioakustik. 286 pp. A. Ziemsen Verlag, Wittenberg Lutherstadt.

A technical review of animal voices and communication by sounds. Biblio. (General)

Thorpe, W. H. 1956 Learning and instinct in animals. viii + 493 pp. Harvard Univ. Press, Cambridge, Mass. A technical, general review of these subjects. Biblio. (2,3)

Thorpe, W. H. 1961 Bird-song. xii + 143 pp. Cambridge Univ. Press, Cambridge, England. A technical review of bird song and its communicative significance. (General)

Thorpe, W. H. and O. L. Zangwill (Eds.) 1961 Current problems in animal behaviour. xiv + 424 pp. Cambridge Univ. Press, Cambridge, England. Technical reports by a number of scientists on research in animal behavior, including some on communication. (General)

Tinbergen, N. 1951 The study of instinct. xii + 228 pp. Oxford Univ. Press, Oxford, England. A nontechnical review of experimental ethological methods in behavioral studies. Biblio. (2,3,7)

Tinbergen, N. 1953 The herring gull's world. A study of the social behaviour of birds. xvi + 255 pp. Collins, London. A nontechnical report of observations and experiments on the life of this species. (General)

Tinbergen, N. 1953 Social behaviour in animals, with special reference to vertebrates. xi + 150 pp. Methuen & Co. Ltd., London. A semipopular, brief review of all aspects of animal social life. (General)

Tinbergen, N. 1960 The evolution of behavior in gulls. Sci. American, 203(6): 118–126; 128; 130. A nontechnical report of observations on social and sexual communication. (4,7,8,9)

Warden, Carl, Thomas N. Jenkins, and Lucien H. Warner 1940 Comparative psychology. 3 Vols. xxxiii + 2136 pp. A technical general review of studies on senses, behavior, and social relations of animals in all aspects. Extensive biblio. (General)

Washburn, S. L. and Irven DeVore 1961 The social life of baboons. Sci. American, 204(6): 62–71. A nontechnical report of field observations on these animals. (3,4,5,6,7,8)

Wheeler, William Morton 1910 Ants, their structure, development and behavior. xxv + 663 pp. Columbia Univ. Press, New

York. A technical review of the biology of ants in all aspects. Biblio. (4,5,8)

Wilson, Edward O. 1963 Pheromones. Sci. American, 208(5): 100–106; 108; 110; 112; 114. A nontechnical review of research on chemical communication signals. (3,4,5,6,9)

Wynne-Edwards, V. C. 1962 Animal dispersion in relation to social behaviour. xi + 653 pp. Hafner Publ. Co., New York. A technical review of the importance of animal behavior and communication in determining distribution of animals. Biblio. (General)

Index Table

of Animal Groups and Systems of Communication

The functions of the communication systems are indicated under each listing as follows: Sp = Species Identification; So = Social Cooperation; Se = Sexual Attraction and Recognition; Co = Courtship and Mating; Pa = Parental Care and Development.

	Tactile	*Chemical*	*Optical*	*Acoustical*
MAMMALS	Sp: 46; Co: 113; Pa: 126	Sp: 42, 50; So: 166; Se: 14, 76, 161; Pa: 126, 132	Sp: 49; So: 16, 57; Pa: 126, 128, 132	Sp: 50; So: 4, 57, 58, 60, 63, 141, 160, 171; Co: 113; Pa: 128
BIRDS	Sp: 46; Co: 112; Pa: 127	Sp: 14, 42	Sp: 15, 41, 48, 51; So: 16, 56; Se: 36, 147, 161; Co: 112; Pa: 35, 122, 125, 126	Sp: 41, 50; So: 4, 18, 35, 36, 56, 57, 58, 59, 63, 135, 140, 141, 147, 160, 169, 173; Se: 86, 162; Co: 112; Pa: 122, 127, 131
REPTILES and AMPHIBIANS		Co: 111	Sp: 48; Se: 111; Co: 111	Se: 19, 35, 84, 139; Co: 111
FISHES		So: 41, 55; Se: 76, 160	Sp: 48; So. 35, 41, 54; Se: 36, 90; Co: 110; Pa: 130	Se: 84, 160; Co: 110

INSECTS	So: 5, 11, 12, 26, 54, 55, 63, 65, 137, 149, 163; Pa: 121, 123	Sp: 42; So: 54, 62, 63, 65, 149; Se: 14, 15, 27, 30, 74, 165; Co: 108, 109; Pa: 124	Sp: 47; So: 41; Se: 15, 88, 137; Co: 108; Pa: 129	Sp: 47, 49; So: 54, 58, 59, 167; Se: 17, 23, 31, 34–36, 77, 89, 137, 139, 144, 146, 149, 154, 167; Co: 108; Pa: 121, 125, 130
SPIDERS and SCORPIONS	Co: 98, 104	Se: 74; Co: 107	Sp: 47; Co: 105	Se: 20; Co: 102, 105
CRUSTACEA	Sp: 46	Sp: 39	Sp: 46; So: 16, 36; Se: 16; Co: 97	Sp: 39; So: 19; Se: 77
OTHER INVERTE-BRATES	So: 12; Co: 93, 95	Sp: 39; Se: 5, 38, 74; Co: 93; Pa: 119, 120	Se: 88; Co: 93	

*Index**

* Italic numbers in the Index refer to the page numbers on which figures appear.

PRINTED IN THE UNITED STATES OF AMERICA

ABOUT THE AUTHORS

Hubert and Mable Frings have been at the University of Hawaii since 1961, where he is Professor of Zoology, teaching general zoology, invertebrate zoology, and advanced courses, and she is Research Associate in Zoology, doing research and bibliographic work on sensory physiology and animal communication. Mable Frings received her B.S. from Pennsylvania State University in 1935, and Hubert Frings received his B.S. from there in 1936. They were married in 1936. Hubert Frings received his M.S. degree from the University of Oklahoma in 1937 and his Ph.D. degree from the University of Minnesota in 1940. From 1940–1947 the Frings were at Monett Junior College, Snead Junior College, West Virginia Wesleyan College, and Gustavus Adolphus College. In 1947 they went to Pennsylvania State University, where Dr. Frings taught zoology and entomology and Mrs. Frings did research on sensory physiology. Hubert and Mable Frings hold memberships in: American Association for the Advancement of Science, Entomological Society of America, American Society of Zoologists, Marine Biological Association of the United Kingdom, Ecological Society of America, Acoustical Society of America, Society of Protozoologists, and American Society of Human Genetics.

THIS BOOK WAS SET IN

ELECTRA AND DEEPDENE TYPES BY

ATLANTIC LINOTYPE COMPANY, INC.

TYPOGRAPHY AND DESIGN

ARE BY THE STAFF OF

BLAISDELL PUBLISHING COMPANY.